SpringerBriefs in Electrical and Computer Engineering

SpringerBriefs present concise summaries of cutting-edge research and practical applications across a wide spectrum of fields. Featuring compact volumes of 50 to 125 pages, the series covers a range of content from professional to academic. Typical topics might include: timely report of state-of-the art analytical techniques, a bridge between new research results, as published in journal articles, and a contextual literature review, a snapshot of a hot or emerging topic, an in-depth case study or clinical example and a presentation of core concepts that students must understand in order to make independent contributions.

More information about this series at http://www.springer.com/series/10059

Ying-Jun Angela Zhang • Congmin Fan
Xiaojun Yuan

Scalable Signal Processing in Cloud Radio Access Networks

Ying-Jun Angela Zhang
Department of Information Engineering
Chinese University of Hong Kong
Shatin, New Territories, Hong Kong

Congmin Fan
Department of Information Engineering
Chinese University of Hong Kong
Shatin, New Territories, Hong Kong

Xiaojun Yuan
Center for Intelligent Networking
and Communications
The University of Electronic Science
and Technology of China
Chengdu, China

ISSN 2191-8112 ISSN 2191-8120 (electronic)
SpringerBriefs in Electrical and Computer Engineering
ISBN 978-3-030-15883-5 ISBN 978-3-030-15884-2 (eBook)
https://doi.org/10.1007/978-3-030-15884-2

Library of Congress Control Number: 2019936444

This Springer imprint is published by the registered company Springer Nature Switzerland AG.
The registered company address is: Gewerbestrasse 11, 6330 Cham, Switzerland

Preface

With centralized processing, cooperative radio, real-time cloud computing, and clean infrastructure, cloud radio access network (C-RAN) is a "future-proof" solution to sustain the mobile data explosion in future wireless networks. The technology holds great potential to accommodate the unprecedented traffic volume that today's wireless cellular system is facing. However, the high density of remote radio heads (RRHs) in C-RANs leads to severe scalability issues in terms of computational and implementation complexities. In this book, we will undertake a comprehensive study on scalable collaborative signal processing for C-RANs, where scalable means (1) the computational complexity and the amount of required information (say, channel state information (CSI)) do not grow rapidly with the network size, and (2) the performance is not substantially degraded, as compared with that of the full-scale cooperation. In particular, this book is divided into the following parts. Firstly, the book proposes a threshold-based channel matrix sparsification approach, with which a vast majority of the channel coefficients can be ignored without substantially compromising the system performance. Secondly, this book proposes a training sequence design scheme for time-multiplexing channel training in C-RANs by exploiting the sparsity of the channel matrix. Thirdly, the book proposes two low-complexity centralized signal detection algorithms based on the sparsified channel matrix. In summary, the central theme of this book is to develop scalable PHY-layer collaborative signal processing schemes by utilizing the near-sparsity of the channel matrix in C-RANs. With this book, we wish to spur new research activities in the following important question: how to leverage the revolutionary architecture of C-RAN to attain unprecedented system capacity at an affordable cost and complexity.

Shatin, New Territories, Hong Kong — Ying-Jun Angela Zhang
Shatin, New Territories, Hong Kong — Congmin Fan
Chengdu, China — Xiaojun Yuan

Acknowledgements

This book becomes a reality with the kind support and help of many individuals in the past few years. We would like to extend our sincere thanks to all of them.

Contents

1 Introduction 1
1.1 Backgrounds 1
1.2 Motivations 2
1.3 Contributions 5
1.4 Organization 6
References 7

2 System Model and Channel Sparsification 9
2.1 System Model 9
2.1.1 System Setup 9
2.1.2 Channel Sparsification 10
2.2 Distance Threshold Analysis 11
2.3 Numerical Results 18
2.3.1 Verification 18
2.3.2 Discussion on the Distance Threshold 19
2.4 Conclusions 20
References 21

3 Scalable Channel Estimation 23
3.1 System Model 24
3.1.1 Training Phase 25
3.1.2 Data Transmission Phase 26
3.2 Problem Formulation 26
3.2.1 Throughput Optimization 26
3.2.2 Local Orthogonality 27
3.2.3 Problem Statement 30
3.2.4 Related Work 31
3.3 Training Sequence Design 32
3.4 Optimal Training Length 33
3.4.1 Graph with Infinite RRHs 34
3.4.2 Asymptotic Behavior of the Training Length 34
3.4.3 Further Discussions 36

3.5 Practical Design 38
3.5.1 Refined Channel Sparsification 38
3.5.2 Numerical Results 39
3.6 Conclusions 43
Appendix 43
References 46

4 Scalable Signal Detection: Dynamic Nested Clustering 49
4.1 System Model and Problem Formulation 49
4.2 Single-Layer Dynamic Nested Clustering 51
4.2.1 RRH Labelling Algorithm 52
4.2.2 Single-Layer DNC 53
4.2.3 Optimizing the Computational Complexity 55
4.2.4 Parallel Computing 57
4.3 Multi-Layer DNC Algorithm 58
4.3.1 Two-Layer DNC Algorithm 60
4.3.2 Optimizing the Computational Complexity 61
4.3.3 Parallel Computing 62
4.4 Numerical Results 63
4.5 Conclusions 65
References 65

5 Scalable Signal Detection: Randomized Gaussian Message Passing 67
5.1 Gaussian Message Passing with Channel Sparsification 67
5.1.1 Bipartite Random Geometric Graph 67
5.1.2 Gaussian Message Passing 69
5.1.3 Related Work 71
5.2 Randomized Gaussian Message Passing with Channel Sparsification 73
5.2.1 Randomized Gaussian Message Passing 73
5.2.2 Numerical Examples 74
5.3 Convergence Analysis 77
5.3.1 Convergence of GMP 77
5.3.2 Convergence of RGMP 79
5.4 Blockwise RGMP and Its Convergence Analysis 82
5.4.1 Blockwise RGMP 82
5.4.2 Convergence Analysis of B-RGMP 83
5.5 Numerical Comparisons 85
5.5.1 Comparison of Convergence 85
5.5.2 Comparison of Convergence Speed 86
5.5.3 Comparison of Performance 88
5.6 Conclusions 89
References 90

6 Conclusions and Future Work 93
6.1 Conclusions 93
6.2 Future Work 94
Reference 96

Index 97

Chapter 1
Introduction

1.1 Backgrounds

Driven by the astounding development of smart phones, mobile applications and the Internet of Things (IoT), traffic demand grows exponentially in current mobile networks. Based on the recent statistics from Cisco [1], the global mobile data traffic has grown 18-fold over the past 5 years and is expected to increase sevenfold by 2021. Moreover, the number of mobile-connected devices, including smartphones, wearable devices, machine-to-machine modules, auto pilot cars, etc., is estimated to be 11.6 billion in 2021, which will be approximated to one and a half times of the world's projected population at that time. The rapid proliferation of mobile devices, coupled with an abundance of new applications, will also bring a large variety of new service requirements. For instance, massive machine type communications require high connection density, videos require very high throughput per connection, auto pilot cars require low latency and ultra high reliability, augmented reality requires both high throughput and low latency, and so on.

To meet the increasing traffic demand and the various service requirements, network infrastructure densification emerges as a promising solution. By deploying more antennas and radio access points, the coverage and capacity of cellular networks are improved by exploiting spatial reuse of limited spectrum. However, there exist formidable challenges to achieve the benefits of network densification based on today's radio access network (RAN) architecture. The main challenges are listed as follows.

- The capital (CAPEX) and operational (OPEX) expenses in terms of initial base station (BS) setup, site support, site rental, and daily operation and maintenance rapidly increase with the density of BSs. This means building more BSs results in significant additional cost.
- The energy consumption (such as the consumption of BS and RAN operation, air condition, and transmission and reception of radio signals) is high. For

Y.-J. A. Zhang et al., *Scalable Signal Processing in Cloud Radio Access Networks*, SpringerBriefs in Electrical and Computer Engineering,
https://doi.org/10.1007/978-3-030-15884-2_1

example, each stand-alone BS requires an on-site equipment room and associated air conditioners, leading to huge energy consumption when the number of BSs increases.
- The spectrum efficiency is low. It is common that the coverage areas of BSs in a dense network are overlapping with each other. This implies severe interference at the cell edges, which may leads to low spectrum efficiency if there is no highly-efficient cooperative interference management, radio resource allocation, and frequency reuse strategies.
- The utilization rate of BSs is low. This is because the average network load is usually much lower than that in peak load. Moreover, each BS is isolated to serve a small area, which means that the processing power and storage resources cannot be shared with other BSs.
- The system upgrade and operation are expensive. To support the new booming mobile applications, mobile operators need to upgrade their network frequently. However, since BSs are built on proprietary platforms, frequent upgrade of a large number of BSs becomes impractical.

To resolve the above-mentioned challenges, a revolutionary wireless network architecture, referred to as the cloud radio access network (C-RAN), has emerged as a promising solution. As shown in Fig. 1.1, a C-RAN consists of three key components: (a) the distributed remote radio heads (RRHs), each deployed with antennas at the remote site of a small cell, (b) a pool of baseband units (BBUs) in a data center cloud, run by high-performance general purpose processors (GPPs) and real-time BS virtualization, and (c) a high-bandwidth, low-latency optical transport network connecting the BBUs and RRHs. The key distinction of a C-RAN from traditional base station (BS) systems is that the RRHs are separated from the baseband processing units, and the latter are migrated to a centralized data center. This keeps the RRHs light-weight, thereby allowing them to be deployed in large numbers of small cells with low cost. Meanwhile, centralized processing opens up new possibilities for high spectrum efficiency and cost reduction through collaborative signal processing, flexible interference management, radio and computing resource management, joint synchronization, etc. Moreover, equipped with an open IT platform, the centralized BBU pool can support inexpensive system upgrade and management. As such, C-RAN has been recognized as a "future proof" architecture that addresses the unprecedented challenges today's wireless cellular system is facing.

1.2 Motivations

The novel architecture of C-RAN brings opportunities to significant system capacity enhancement and cost reduction. The exciting opportunities, however, come with unique technical challenges, among which one is particular outstanding. That is, how to leverage the enormous centralized baseband-processing power to attain

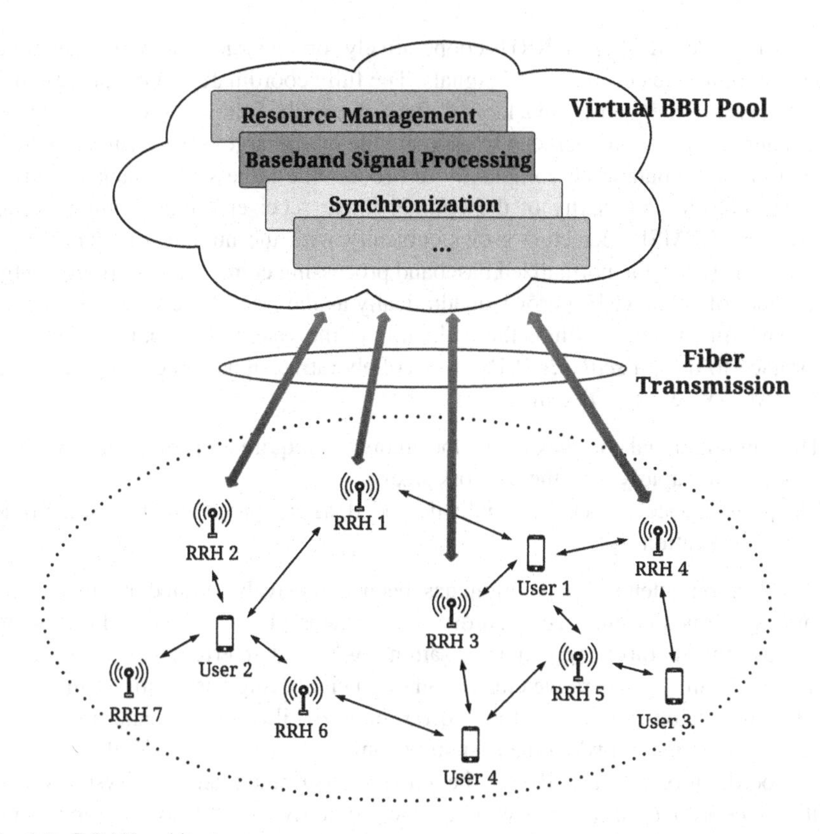

Fig. 1.1 C-RAN architecture

unprecedented system capacity at affordable cost and complexity. In a C-RAN, the virtual BSs work together in a large physical BBU pool, and thus are allowed to easily share the signalling, traffic data, and channel state information (CSI) of active users in the system. On one hand, this enables tight BS coordination, including joint signal processing, scheduling, radio resource management, and load balancing, so as to greatly enhance the system capacity. On the other hand, the complexity and cost of tight coordination of all BSs may increase substantially when the network size becomes large. Indeed, the preliminary C-RAN technology can already support around 10 km separation between the BBU pool and RRHs, covering 10–1000 RRH sites [2]. It is not hard to imagine that the size of the network will grow even larger with the advances of radio over fiber (RoF) and data-center technology. With such a large network size, the cost of full-scale coordination becomes prohibitively high. Therefore, it is critical to develop collaborative network design schemes with low computational and implementational complexities in C-RANs.

In this book, we focus on PHY-layer collaborative signal processing in C-RANs. Due to centralized baseband processing, RRHs in a C-RAN can be viewed as a large-scale distributed antenna system. Theoretically speaking, the highest system capacity is achieved when all RRHs cooperatively form a large-scale virtual antenna array that jointly detects the users' signals. The fully coordinated signal processing, however, is extremely costly in a large C-RAN due to the following two reasons: (a) high channel estimation overhead to estimate the entire large-scale channel matrix, and (b) high computational complexity to process the large-scale channel matrix. For instance, the complexity of the optimal linear receiver (i.e., minimum mean square error (MMSE) detector) grows cubically with the number of users/RRHs [3]. In other words, the normalized baseband processing complexity (normalized by the number of users/RRHs) grows quadratically as the size of the system becomes large. This fundamentally limits the scalability of the system. Hence, it is of utmost importance to design *scalable* PHY-layer collaborative signal processing schemes for C-RANs, where *scalable* means:

1. The computational complexity and the amount of required information (say, CSI) do not grow rapidly with the network size
2. The performance is not substantially degraded, as compared with that of the full-scale cooperation.

Base station/antenna cooperation has been extensively studied in distributed antenna systems and multi-cell coordination systems [4–14]. Most of the existing work has studied throughput maximization [5–7] and interference management [8–14] by forming cooperative clusters among neighboring base stations/antennas. Limited discussions have been focused on the scalability of baseband-processing and channel-estimation when the system becomes extremely large. With the large-scale coordination of a C-RAN, the current distributed antenna systems and multi-cell coordination schemes will become prohibitively expensive to implement. To partially address the problem, a recent work by Shi et al. [15] proposed a low-complexity coordinated beamforming algorithm in C-RANs. Even though the simulation results show that the proposed algorithm can significantly reduce the computation time, [15] does not discuss how the complexity scales with the network size. Moreover, perfect knowledge of the entire channel matrix is required for beamforming, which is impractical for large-scale C-RAN.

In this book, we propose to exploit the near-sparsity of large C-RAN channel matrices to design scalable collaborative signal processing algorithms. In a C-RAN, users and RRHs are scattered over a large area. Due to the propagation attenuation of electromagnetic waves, an RRH usually receives relatively strong signals from only a small number of nearby users. Moreover, the transmission delay prevents the signals from far-away users to be processed. Intuitively, ignoring the signals from far-away users in general does not cause much performance loss. This implies a vast majority of signals over the transmission links can be ignored (i.e., the channel matrix can be significantly sparsified) if we can tolerate a small degradation in the system performance. The high sparsity of the channel matrix can thereby be utilized to reduce the computational complexity and the channel estimation overhead. The detailed contributions of the book will be presented in next section.

1.3 Contributions

In this book, we endeavour to design a C-RAN baseband processing solution that can enjoy the advantage of full-scale RRH coordination, while the amount of required CSI (or the amount of channel estimation overhead) and the complexity are kept at a tolerable level for a network with very large size. The key contributions of this book are summarized into the following three aspects.

- **Channel Matrix Sparsification:** With rigorous analysis, we show that without causing noticeable performance loss, the signals can be detected by processing a sparsified channel matrix instead of the full channel matrix. In particular, we propose a threshold-based channel matrix sparsification method, in which matrix entries are discarded if the corresponding link length (or large-scale fading in general) exceeds a certain threshold. A closed-form expression is derived to relate the threshold to the signal-to-interference-and-noise ratio (SINR) loss due to matrix sparsification. The result shows that for reasonably large networks, a vast majority of the channel coefficients can be ignored if we can tolerate a very small percentage of SINR loss. As shown in this book, this result will not only lead to great reduction of the channel estimation overhead, but also open up the possibility of design low-complexity signal detection algorithms.
- **Scalable Channel Estimation:** We investigate how the sparsity of the channel matrix will affect the channel estimation. Specifically, we consider the design of training sequences for time-multiplexed channel training in C-RANs. Based on the channel sparsification approach, we introduce the notion of *local orthogonality*, in which the training sequences of the users in the neighborhood of an RRH are required to be orthogonal to each other. The training design problem is then formulated as to find the minimum training length that preserves local orthogonality. This problem can be recast as a vertex-coloring problem, and the existing vertex-coloring algorithms [16, 17] can be applied to solve the problem. Further, we analyze the minimum training length as a function of the network size. Based on the theory of random geometric graph, we show that the training length is $O(\ln K)$ almost surely, where K is the number of users. This guarantees a scalable training-based C-RAN design, i.e., the proposed training design can be applied to a large-size C-RAN system satisfying local orthogonality at the cost of a moderate training length.
- **Scalable Signal Detection:** By exploiting the high sparsity of the channel matrix, we propose two low-complexity MMSE detection algorithms for C-RANs. The first algorithm is the dynamic nested clustering (DNC) algorithm. By skillfully indexing the RRHs, the sparsified channel matrix has a (nested) doubly bordered block diagonal (DBBD) structure, in which the diagonal blocks can be interpreted as clusters (or sub-networks) that are processed independently. Different clusters are coordinated by the cut-node block and border blocks that capture the interference among clusters. As such, the baseband-processing complexity is dominated by the size of the clusters instead of the entire C-RAN network. Thanks to the centralized BBU pool of C-RAN, the DNC algorithm is

amenable for different processing implementations through adjusting the size and the number of the clusters. We design different clustering strategies for both serial processing and parallel processing to minimize the computational complexity and the computation time, respectively. With the proposed DNC algorithm, we show that the computational complexity (i.e., the computation time with serial processing) for the optimal linear detector is significantly reduced from $O(N^3)$ to $O(N^2)$, where N is the number of RRHs in the C-RAN. Moreover, the proposed DNC algorithm is also amenable to parallel processing, which further reduces the computation time to $O(N^{\frac{42}{23}})$. The second algorithm is randomized Gaussian message passing. We formulate the signal detection in C-RAN as an inference problem over a bipartite random geometric graph. By passing messages among neighboring nodes, message passing (a.k.a. belief propagation) provides an efficient way to solve the inference problem over a sparse graph. Traditional message-passing algorithms are not guaranteed to converge, because the corresponding bipartite random geometric graph is locally dense and contains many short loops. We propose a randomized Gaussian message passing (RGMP) algorithm to improve the convergence. Instead of exchanging messages simultaneously or in a fixed order, we propose to exchange messages asynchronously in a random order. The proposed RGMP algorithm demonstrates significantly better convergence performance than conventional message passing. In addition, we generalize the RGMP algorithm to a blockwise RGMP (B-RGMP) algorithm, which allows parallel implementation. The average computation time of B-RGMP remains constant when the network size increases.

1.4 Organization

The book is organized as follows. Chapter 2 describes the system model for Chaps. 2–5. Chapter 2 also proposes a threshold-based channel sparsification approach, and analyzes the effect of channel sparsification on the SINR performance of uplink MMSE detection in C-RANs. Chapter 3 considers training-based channel estimation for C-RANs. A training design scheme based on graph coloring is proposed. Chapters 4 and 5 propose two low-complexity signal detection algorithms. In particular, in Chap. 4, a dynamic nested clustering (DNC) algorithm is proposed based on a unified theoretical framework for dynamic clustering by exploiting the near-sparsity of large C-RAN channel matrices. In Chap. 5, we present a randomzied Gaussian message passing (RGMP) algorithm, whose average computational time remains constant when the network size increases. Finally, Chap. 6 summarizes the key conclusions in this book, and discusses several potential directions of future work.

References

1. *Cisco visual networking index: Global mobile data traffic forecast update, 2016–2021*. Cisco, Technical Report, 2017.
2. C. Mobile, "C-RAN: The road towards green RAN," *White Paper, ver*, vol. 2, 2011.
3. M. Tuchler, A. C. Singer, and R. Koetter, "Minimum mean squared error equalization using a priori information," *IEEE Transactions on Signal Processing*, vol. 50, no. 3, pp. 673–683, 2002.
4. A. Abdelnasser, E. Hossain, and D. I. Kim, "Clustering and resource allocation for dense femtocells in a two-tier cellular OFDMA network," *IEEE Transactions on Wireless Communications*, vol. 13, no. 3, pp. 1628–1641, 2014.
5. R. Zhang, "Cooperative multi-cell block diagonalization with per-base-station power constraints," *IEEE Journal on Selected Areas in Communications*, vol. 28, no. 9, pp. 1435–1445, 2010.
6. J. Zhang, R. Chen, J. G. Andrews, A. Ghosh, and R. W. Heath, "Networked MIMO with clustered linear precoding," *IEEE Transactions on Wireless Communications*, vol. 8, no. 4, 2009.
7. A. Liu and V. K. Lau, "Joint power and antenna selection optimization in large cloud radio access networks," *IEEE Transactions on Signal Processing*, vol. 62, no. 5, pp. 1319–1328, 2014.
8. S. Samarakoon, M. Bennis, W. Saad, and M. Latva-aho, "Dynamic clustering and on/off strategies for wireless small cell networks," *IEEE Transactions on Wireless Communications*, vol. 15, no. 3, pp. 2164–2178, 2016.
9. M. Hong, Z. Xu, M. Razaviyayn, and Z.-Q. Luo, "Joint user grouping and linear virtual beamforming: Complexity, algorithms and approximation bounds," *IEEE Journal on Selected Areas in Communications*, vol. 31, no. 10, 2013.
10. A. Liu and V. Lau, "Hierarchical interference mitigation for massive MIMO cellular networks," *IEEE Transactions on Signal Processing*, vol. 62, no. 18, pp. 4786–4797, 2014.
11. J. Gong, S. Zhou, Z. Niu, L. Geng, and M. Zheng, "Joint scheduling and dynamic clustering in downlink cellular networks," in *Proc. of IEEE Global Communications Conference (GLOBECOM)*, 2011, pp. 1–5.
12. N. Lee, D. Morales-Jimenez, A. Lozano, and R. W. Heath, "Spectral efficiency of dynamic coordinated beamforming: A stochastic geometry approach," *IEEE Transactions on Wireless Communications*, vol. 14, no. 1, pp. 230–241, 2015.
13. Y. Shi, J. Zhang, and K. B. Letaief, "Optimal stochastic coordinated beamforming for wireless cooperative networks with CSI uncertainty," *IEEE Transactions on Signal Processing*, vol. 63, no. 4, pp. 960–973, 2015.
14. L. Dai and B. Bai, "Optimal decomposition for large-scale infrastructure-based wireless networks," *IEEE Transactions on Wireless Communications*, vol. 16, no. 8, pp. 4956–4969, 2017.
15. Y. Shi, J. Zhang, and K. B. Letaief, "Scalable coordinated beamforming for dense wireless cooperative networks," in *Proc. of IEEE Global Communications Conference (GLOBECOM)*, 2014, pp. 3603–3608.
16. D. Brélaz, "New methods to color the vertices of a graph," *Communications of the ACM*, vol. 22, no. 4, pp. 251–256, 1979.
17. E. Malaguti, M. Monaci, and P. Toth, "A metaheuristic approach for the vertex coloring problem," *INFORMS Journal on Computing*, vol. 20, no. 2, pp. 302–316, 2008.

Chapter 2
System Model and Channel Sparsification

In C-RAN, only a small fraction of the entries in the channel matrix have reasonably large amplitudes, because a user is only close to a small number of RRHs in its neighborhood, and vice versa. Thus, ignoring the small entries in the channel matrix would significantly sparsify the matrix, which can potentially lead to significant reduction in the computational complexity and channel estimation overhead. The question is to what extent can the channel matrix be sparsified without substantially compromising the system performance. In this chapter, we attempt to address this question. In particular, we propose a threshold-based channel matrix sparsification method, where the matrix entries are ignored according to the distance between the users and RRHs. We derive a closed-form expression describing the relationship between the threshold and the SINR loss due to channel spasification. The analysis serves as a convenient guideline to set the threshold subject to a tolerable SINR loss. The material in this chapter is mainly based on [1].

2.1 System Model

2.1.1 System Setup

We consider the uplink transmission of a C-RAN with N single-antenna RRHs, and K single-antenna mobile users uniformly located over the entire coverage area. The received signal vector $\mathbf{y} \in \mathbb{C}^{N\times 1}$ at the RRHs is

$$\mathbf{y} = \mathbf{H}\mathbf{P}^{\frac{1}{2}}\mathbf{x} + \mathbf{n}, \tag{2.1}$$

where $\mathbf{H} \in \mathbb{C}^{N\times K}$ denotes the channel matrix, with the (n, k)th entry $H_{n,k}$, being the channel coefficient between the kth user and the nth RRH. $\mathbf{P} \in \mathbb{R}^{N\times N}$ is

Y.-J. A. Zhang et al., *Scalable Signal Processing in Cloud Radio Access Networks*, SpringerBriefs in Electrical and Computer Engineering,
https://doi.org/10.1007/978-3-030-15884-2_2

a diagonal matrix with the kth diagonal entry P_k being the transmitting power allocated to user k. $\mathbf{x} \in \mathbb{C}^{K\times 1}$ is the vector of the transmitted signal from the K users and $\mathbf{n} \sim \mathcal{CN}(\mathbf{0}, N_0\mathbf{I})$ is the vector of noise received by RRHs. The transmit signals are assumed to follow an independent complex Gaussian distribution with unit variance, i.e. $E[\mathbf{x}\mathbf{x}^{\mathrm{H}}] = \mathbf{I}$. Specifically, $H_{n,k} = \gamma_{n,k} d_{n,k}^{-\frac{\zeta}{2}}$, where $\gamma_{n,k}$ is the i.i.d. Rayleigh fading coefficient with zero mean and variance 1, $d_{n,k}$ is the distance between the nth RRH and the kth user, and ζ is the path loss exponent. Then, $d_{n,k}^{-\zeta}$ is the path loss from the kth user to the nth RRH.

Without loss of generality, let us consider user k. The receive detection vector is

$$\mathbf{v}_k = P_k^{\frac{1}{2}} (\mathbf{H}\mathbf{P}\mathbf{H}^{\mathrm{H}} + N_0\mathbf{I})^{-1}\mathbf{h}_k, \tag{2.2}$$

and the decision statistics of x_k is

$$\widehat{x}_k = \mathbf{v}_k^{\mathrm{H}}\mathbf{h}_k P_k^{\frac{1}{2}} x_k + \mathbf{v}_k^{\mathrm{H}} \sum_{j\neq k} \mathbf{h}_j P_j^{\frac{1}{2}} x_j + \mathbf{v}_k^{\mathrm{H}}\mathbf{n}, \tag{2.3}$$

where $\mathbf{h}_k \in \mathbb{C}^{N\times 1}$ is the kth column of the channel matrix $\mathbf{H}$. The SINR for user k is

$$\mathrm{SINR}_k = \frac{P_k |\mathbf{v}_k^{\mathrm{H}}\mathbf{h}_k|^2}{\sum_{j\neq k} P_j |\mathbf{v}_k^{\mathrm{H}}\mathbf{h}_j|^2 + N_0 \mathbf{v}_k^{\mathrm{H}}\mathbf{v}_k}. \tag{2.4}$$

Notice that to calculate the detection vector $\mathbf{v}_k$, the full channel matrix $\mathbf{H}$ needs to be acquired and processed. In particular, the complexity of calculating inverse of $\mathbf{H}\mathbf{P}\mathbf{H}^{\mathrm{H}} + N_0\mathbf{I}$ is as high as $O(N^3)$ by using Gaussian elimination. The polynomial increase in computational complexity causes a serious scalability problem. That is, the average computational complexity per user increases quadratically with the network size, if K scales in the same order as N, rendering the full-scale RRH cooperation very costly in large-scale C-RANs. To address the issue, we will argue in the next subsection that most entries in $\mathbf{H}$ are insignificant, and thus can be ignored. The effect of sparsifying matrix $\mathbf{H}$ by ignoring insignificant entries will be quantified analytically in Sect. 2.2.

2.1.2 Channel Sparsification

Since the RRHs and users are distributed over a large area, an RRH can only receive reasonably strong signals from a small number of nearby users, and vice versa. Thus, ignoring the small entries in $\mathbf{H}$ would significantly sparsify the matrix, hopefully with a negligible loss in system performance. In this book, we propose to ignore the entries of $\mathbf{H}$ based on the distance of links. In other words, the entry $H_{n,k}$ is set to 0 when the link length $d_{n,k}$ is larger than a threshold d_0. The resulting sparsified channel matrix, denoted by $\widehat{\mathbf{H}}$, is given by

$$\widehat{H}_{n,k} = \begin{cases} H_{n,k}, & d_{n,k} < d_0 \\ 0, & \text{otherwise.} \end{cases} \tag{2.5}$$

Note that we propose to sparsify the channel matrix based on the link distance instead of the actual values of entries, which are affected by both the link distance and fast channel fading. In practice, link distances vary much more slowly than fast channel fading. The distance-threshold-based approach leads to a relatively stable structure of $\widehat{\mathbf{H}}$. The received signal $\mathbf{y}$ can then be represented as

$$\mathbf{y} = \widehat{\mathbf{H}}\mathbf{P}^{\frac{1}{2}}\mathbf{x} + \widetilde{\mathbf{H}}\mathbf{P}^{\frac{1}{2}}\mathbf{x} + \mathbf{n}, \tag{2.6}$$

where $\widetilde{\mathbf{H}} = \mathbf{H} - \widehat{\mathbf{H}}$. Treating the interference signal $\widetilde{\mathbf{H}}\mathbf{P}^{\frac{1}{2}}\mathbf{x}$ as noise, the detection vector becomes

$$\widehat{\mathbf{v}}_k = P_k^{\frac{1}{2}} \left(\widehat{\mathbf{H}}\mathbf{P}\widehat{\mathbf{H}}^{\mathrm{H}} + \boldsymbol{\Gamma} + N_0\mathbf{I} \right)^{-1} \widehat{\mathbf{h}}_k, \tag{2.7}$$

where $\widehat{\mathbf{h}}_k$ is the kth column of $\widehat{\mathbf{H}}$, and

$$\boldsymbol{\Gamma} = \mathrm{E}\left[\sum_{j \neq k} P_j \left(\widetilde{\mathbf{h}}_j \widehat{\mathbf{h}}_j^{\mathrm{H}} + \widehat{\mathbf{h}}_j \widetilde{\mathbf{h}}_j^{\mathrm{H}} + \widetilde{\mathbf{h}}_j \widetilde{\mathbf{h}}_j^{\mathrm{H}} \right) \right], \tag{2.8}$$

is the variance of the ignored signal $\widetilde{\mathbf{H}}\mathbf{P}^{\frac{1}{2}}\mathbf{x}$.

With this, the SINR becomes

$$\widehat{\mathrm{SINR}}_k(d_0) = \frac{P_k |\widehat{\mathbf{v}}_k^{\mathrm{H}} \mathbf{h}_k|^2}{\sum_{j \neq k} P_j |\widehat{\mathbf{v}}_k^{\mathrm{H}} \mathbf{h}_j|^2 + N_0 \widehat{\mathbf{v}}_k^{\mathrm{H}} \widehat{\mathbf{v}}_k}. \tag{2.9}$$

Notice that when the distance threshold d_0 is small, the matrix $\widehat{\mathbf{H}}$ can be very sparse, leading to a significant reduction in channel estimation overhead and processing complexity. The key question is: how small d_0 can be without significantly affecting the system performance. This question will be answered in the next section.

2.2 Distance Threshold Analysis

In this section, we show by rigorous analysis how to set the distance threshold d_0 if a high percentage of full SINR is to be achieved. Specifically, we wish to set d_0, such that the SINR ratio, defined as

$$\rho(d_0) = \frac{\mathrm{E}[\widehat{\mathrm{SINR}}_k(d_0)]}{\mathrm{E}[\mathrm{SINR}_k]} \tag{2.10}$$

is larger than a prescribed ρ^*, where the two expectations are taken over entries in $\mathbf{H}$, which are affected by path loss and Rayleigh fading.

We first introduce two approximations that make our analysis tractable.

Approximation 1 The distances $d_{n,k}$ for all n, k are mutually independent.

As shown in Fig. 2.1a, we plot the SINR ratio for systems with and without Approximation 1. The system area is assumed to be a circle with radius 5 km. The figure shows that the gap between the SINR ratios is negligible, which validates the independence approximation.

Approximation 2 Given the distance threshold d_0, the matrices $\widehat{\mathbf{H}}$ and $\widetilde{\mathbf{H}}$ are mutually independent.

Note that $\mathrm{E}[\widehat{\mathbf{H}}\widetilde{\mathbf{H}}^{\mathrm{H}}] = \mathrm{E}[\widetilde{\mathbf{H}}\widehat{\mathbf{H}}^{\mathrm{H}}] = \mathbf{0}$, implying that $\widehat{\mathbf{H}}$ and $\widetilde{\mathbf{H}}$ are uncorrelated. With the independence approximation, the equality

$$\mathrm{E}_{\mathbf{H}}[\widehat{\mathrm{SINR}}_k(d_0)] = \mathrm{E}_{\widehat{\mathbf{H}}}\left[\mathrm{E}_{\widetilde{\mathbf{H}}}[\widehat{\mathrm{SINR}}_k(d_0)]\right] \tag{2.11}$$

holds. This approximation is verified in Fig. 2.1b, which shows that the gap between the simulated SINR ratios with and without Approximation 2 is marginal.

Based on the above two approximations, we see that $\mathbf{\Gamma} = N_1\mathbf{I}$, where $N_1 = \mathrm{E}[\sum_{j=1}^{K} P_j |\widetilde{h}_{n,j}|^2]$. We are now ready to present a lower bound of $\rho(d_0)$ as follows.

Theorem 2.1 *Given a distance threshold d_0, a lower bound of SINR ratio $\rho(d_0)$ is given by*

$$\rho(d_0) \geq \underline{\rho(d_0)} \triangleq \frac{\widehat{\mu} N_0}{\mu\left((\mu - \widehat{\mu})\sum_{j\neq k} P_k + N_0\right)}, \tag{2.12}$$

where $\widehat{\mu} = \int_{x=0}^{d_0} x^{-\zeta} f(x)dx$ and $\mu = \int_{x=0}^{\infty} x^{-\zeta} f(x)dx$, respectively, and $f(x)$ is the probability density function of the distances between RRHs and users. When each user transmits with the same amount of power P, the lower bound is simplified as

$$\underline{\rho(d_0)} = \frac{\widehat{\mu} N_0}{\mu\left(P(\mu - \widehat{\mu})(K-1) + N_0\right)}. \tag{2.13}$$

Proof From (2.9), we have

$$\mathrm{E}\left[\widehat{\mathrm{SINR}}_k(d_0)\right] \tag{2.14a}$$

$$= \mathrm{E}\left[\frac{P_k\left(|\widehat{\mathbf{v}}_k^{\mathrm{H}}\widehat{\mathbf{h}}_k|^2 + \widehat{\mathbf{v}}_k^{\mathrm{H}}\left(\widehat{\mathbf{h}}_k\widetilde{\mathbf{h}}_k^{\mathrm{H}} + \widetilde{\mathbf{h}}_k\widehat{\mathbf{h}}_k^{\mathrm{H}} + \widetilde{\mathbf{h}}_k\widetilde{\mathbf{h}}_k^{\mathrm{H}}\right)\widehat{\mathbf{v}}_k\right)}{\widehat{\mathbf{v}}_k^{\mathrm{H}}\left(\sum_{j\neq k} P_j\mathbf{h}_j\mathbf{h}_j^{\mathrm{H}} + N_0\mathbf{I}\right)\widehat{\mathbf{v}}_k}\right] \tag{2.14b}$$

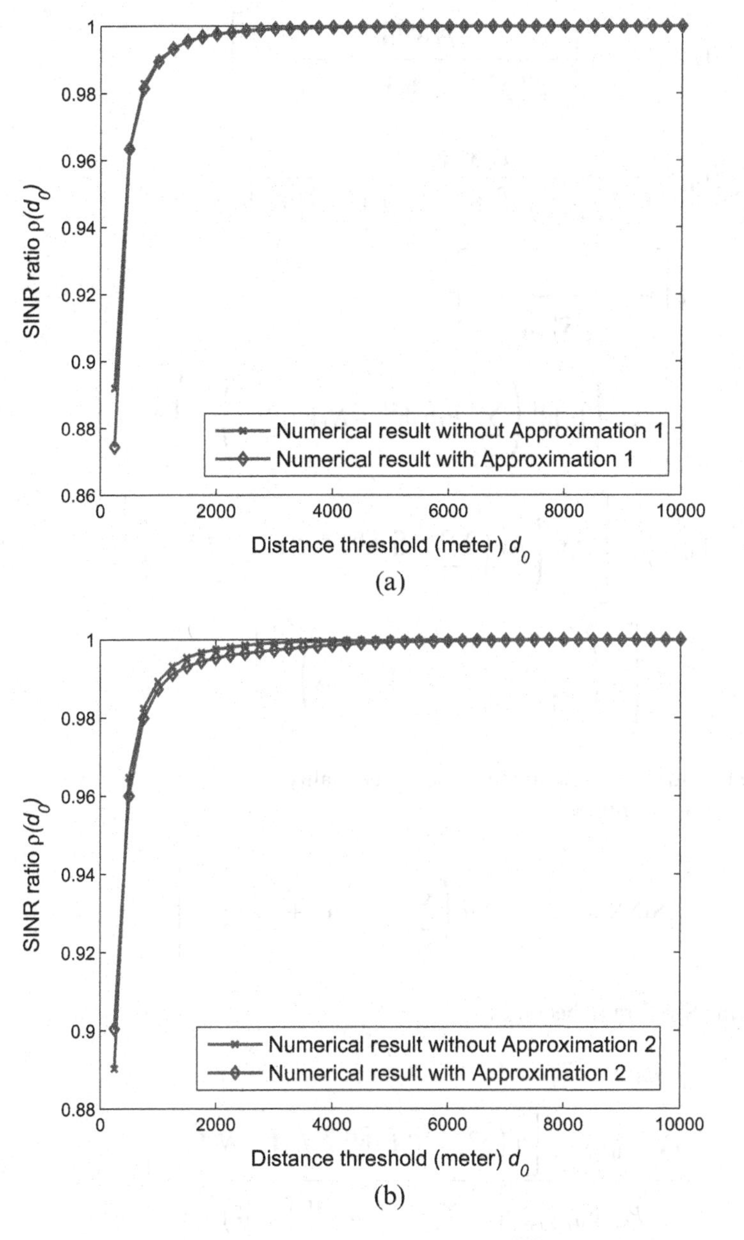

Fig. 2.1 Average SINR ratio vs distance threshold when $N = 1000$, $K = 600$. (**a**) Approximation 1. (**b**) Approximation 2

$$\geq \mathrm{E}_{\widehat{\mathbf{H}}}\mathrm{E}_{\widetilde{\mathbf{h}}_j,\forall j\neq k}\left[\frac{P_k|\widehat{\mathbf{v}}_k^{\mathrm{H}}\widehat{\mathbf{h}}_k|^2}{\widehat{\mathbf{v}}_k^{\mathrm{H}}\left(\sum_{j\neq k}P_j\mathbf{h}_j\mathbf{h}_j^{\mathrm{H}}+N_0\mathbf{I}\right)\widehat{\mathbf{v}}_k}\right] \tag{2.14c}$$

$$\geq \mathrm{E}_{\widehat{\mathbf{H}}}\left[\frac{P_k|\widehat{\mathbf{v}}_k^{\mathrm{H}}\widehat{\mathbf{h}}_k|^2}{\widehat{\mathbf{v}}_k^{\mathrm{H}}\left(\sum_{j\neq k}P_j\widehat{\mathbf{h}}_j\widehat{\mathbf{h}}_j^{\mathrm{H}}+N_1\mathbf{I}+N_0\mathbf{I}\right)\widehat{\mathbf{v}}_k}\right] \tag{2.14d}$$

$$=\mathrm{E}_{\mathbf{H}}\left[\frac{1}{1-P_k^{\frac{1}{2}}\widehat{\mathbf{v}}_k^{\mathrm{H}}\widehat{\mathbf{h}}_k}-1\right] \tag{2.14e}$$

$$=\mathrm{E}_{\widehat{\mathbf{H}}}\left[P_k\mathrm{tr}\left(\widehat{\mathbf{h}}_k\widehat{\mathbf{h}}_k^{\mathrm{H}}\left(\sum_{j\neq k}P_j\widehat{\mathbf{h}}_j\widehat{\mathbf{h}}_j^{\mathrm{H}}+N_1\mathbf{I}+N_0\mathbf{I}\right)^{-1}\right)\right] \tag{2.14f}$$

$$=\mathrm{E}_{\widehat{\mathbf{h}}_j,\forall j\neq k}\left[P_k\mathrm{tr}\left(\widehat{\mu}\mathbf{I}\left(\sum_{j\neq k}P_j\widehat{\mathbf{h}}_j\widehat{\mathbf{h}}_j^{\mathrm{H}}+N_1\mathbf{I}+N_0\mathbf{I}\right)^{-1}\right)\right] \tag{2.14g}$$

$$=P_k\widehat{\mu}\mathrm{E}\left[\mathrm{tr}\left(\sum_{j\neq k}P_j\widehat{\mathbf{h}}_j\widehat{\mathbf{h}}_j^{\mathrm{H}}+N_1\mathbf{I}+N_0\mathbf{I}\right)^{-1}\right], \tag{2.14h}$$

where (2.14d) holds due to the Jensen's inequality.

Likewise, we obtain

$$\mathrm{E}\left[\mathrm{SINR}_k\right]=P_k\mu\mathrm{E}\left[\mathrm{tr}\left(\sum_{j\neq k}P_j\mathbf{h}_j\mathbf{h}_j^{\mathrm{H}}+N_0\mathbf{I}\right)^{-1}\right]. \tag{2.15}$$

Then, the SINR ratio becomes

$$\rho(d_0) \tag{2.16a}$$

$$\geq\frac{P_k\widehat{\mu}\mathrm{E}_{\widehat{\mathbf{h}}_j,\forall j\neq k}\left[\mathrm{tr}\left(\sum_{j\neq k}P_j\widehat{\mathbf{h}}_j\widehat{\mathbf{h}}_j^{\mathrm{H}}+N_1\mathbf{I}+N_0\mathbf{I}\right)^{-1}\right]}{P_k\mu\mathrm{E}_{\mathbf{h}_j,\forall j\neq k}\left[\mathrm{tr}\left(\sum_{j\neq k}P_j\mathbf{h}_j\mathbf{h}_j^{\mathrm{H}}+N_0\mathbf{I}\right)^{-1}\right]} \tag{2.16b}$$

$$\geq\frac{P_k\widehat{\mu}\mathrm{E}_{\mathbf{h}_j,\forall j\neq k}\left[\mathrm{tr}(\sum_{j\neq k}P_j\mathbf{h}_j\mathbf{h}_j^{\mathrm{H}}+N_1\mathbf{I}+N_0\mathbf{I})^{-1}\right]}{P_k\mu\mathrm{E}_{\mathbf{h}_j,\forall j\neq k}\left[\mathrm{tr}\left(\sum_{j\neq k}P_j\mathbf{h}_j\mathbf{h}_j^{\mathrm{H}}+N_0\mathbf{I}\right)^{-1}\right]} \tag{2.16c}$$

$$=\frac{\hat{\mu}}{\mu}\mathrm{E}\left[\frac{\sum_{i=1}^{N}\frac{1}{\lambda_i+N_1+N_0}}{\sum_{i=1}^{N}\frac{1}{\lambda_i+N_0}}\right] \tag{2.16d}$$

$$\geq\frac{\hat{\mu}N_0}{\mu(N_1+N_0)}, \tag{2.16e}$$

where $\lambda_1, \lambda_2, \cdots, \lambda_N$ are the eigenvalues of the positive semidefinite matrix $\sum_{j\neq k} P_j \mathbf{h}_j \mathbf{h}_j^{\mathrm{H}}$. (2.16e) holds since $N_1 \geq 0$, $N_0 \geq 0$ and $\lambda_i \geq 0, \forall i$.

Substituting $N_1 = (\mu - \widehat{\mu}) \sum_{j\neq k} P_k$ into (2.16e), we have

$$\rho(d_0) \geq \frac{\widehat{\mu}N_0}{\mu\left((\mu-\widehat{\mu})\sum_{j\neq k} P_k + N_0\right)}. \tag{2.17}$$

When the users transmit equal power, i.e., $P_1 = P_2 = \cdots = P_K = P$, we have

$$\rho(d_0) \geq \frac{\widehat{\mu}N_0}{\mu\left((\mu-\widehat{\mu})(K-1)P + N_0\right)}. \tag{2.18}$$

We note that $\rho(d_0)$ depends on the probability distribution of the distances between mobile users and RRHs. In [2], distance distributions are derived for different network area shapes, such as circle, square, and rectangle. Consider, for example, a circular network area with radius r. In this case, the distance distribution between two random points is given by [2]

$$f(x,r) = \begin{cases} F(r_0,r) = \delta(x-r_0)\int_0^{r_0} f_1(y,r)\mathrm{d}y, & x \leq r_0, \\ f_1(x,r), & r_0 < x < 2r, \end{cases} \tag{2.19}$$

where r_0 is the minimum distance between an RRH and a user, and

$$f_1(x,r) = \frac{2x}{r^2}\left(\frac{2}{\pi}\arccos\left(\frac{x}{2r}\right) - \frac{x}{\pi r}\sqrt{1-\frac{x^2}{4r^2}}\right). \tag{2.20}$$

Substituting (2.19) into (2.12) and (2.13), we obtain the relation between d_0 and the SINR requirement ρ^*:

Theorem 2.2 *When the distance threshold d_0 satisfies the K inequalities:*

$$\begin{aligned} &N_0 \int_{x=r_0}^{d_0} x^{-\zeta} f(x,r)dx \\ &\geq \rho^* \sum_{j\neq k} P_k \left(\int_{x=d_0}^{2r} x^{-\zeta} f(x,r)dx\right)\left(\int_{x=r_0}^{2r} x^{-\zeta} f(x,r)dx\right), \quad \forall k, \end{aligned} \tag{2.21}$$

where $f(x,r)$ is given in (2.19), an SINR ratio no less than ρ^* can be achieved. When each user transmits with the same amount of power P, d_0 is simplified to the solution of

$$\begin{aligned}&N_0\int_{x=r_0}^{d_0}x^{-\zeta}f(x,r)dx\\&=\rho^*P\left(K-1\right)\left(\int_{x=d_0}^{2r}x^{-\zeta}f(x,r)dx\right)\left(\int_{x=r_0}^{2r}x^{-\zeta}f(x,r)dx\right).\end{aligned}\tag{2.22}$$

When the network size goes to infinity ($r\rightarrow\infty$), the limit of the solution to (2.22) is

$$\lim_{r\rightarrow\infty}d_0=\left(\frac{2N_0(\zeta-2)+2\zeta r_0^{\zeta-2}\rho^*\pi\beta_K P}{\zeta r_0^{2-\zeta}N_0(1-\rho^*)(\zeta-2)}\right)^{\frac{1}{\zeta-2}},\tag{2.23}$$

where $\beta_K=\frac{K}{\pi r^2}$ is the density of users in the network.

Proof We directly obtain (2.21) and (2.22) by substituting (2.19) into (2.12) and (2.13). Then, we focus on the proof of (2.23), i.e., the limit of the distance threshold d_0 as the network radius r goes to infinity. In practice, the path loss exponent ζ is always greater than 2. We further assume that ζ is not an odd integer.[1] Then, the antiderivative of $x^{-\zeta}f_1(x,r)$ denoted as $G(x,r)$ is given below:

$$G(x,r)=\frac{x^{3-\zeta}}{r^3}G_1(x,r)-\frac{x^{2-\zeta}}{r^2}G_2(x,r),\tag{2.24}$$

where $G_2(x,r)=\frac{4}{(\zeta-2)\pi}\cos^{-1}(\frac{x}{2r})$ and

$$\begin{aligned}&G_1(x,r)\\&=\frac{2}{(\zeta-2)(\zeta-3)\pi}\left[(\zeta-2)_2F_1(-\frac{1}{2},\frac{3-\zeta}{2};\frac{5-\zeta}{2};\frac{x^2}{4r^2})\right.\\&\left.+{}_2F_1(\frac{1}{2},\frac{3-\zeta}{2};\frac{5-\zeta}{2};\frac{x^2}{4r^2})\right],\end{aligned}\tag{2.25}$$

with ${}_2F_1(a,b;c;z)=\sum_{n=0}^{\infty}\frac{(a)_n(b)_n}{(c)_n}\frac{z^n}{n!}$ being a hypergeometric function. Here, $(a)_n=a(a+1)\cdots(a+n-1)$ is a Pochhammer symbol. Then,

$$\int_a^b x^{-\zeta}f(x,r)dx=\begin{cases}r_0^{-\zeta}F(r_0,r)+G(b,r)-G(r_0,r) & a=r_0,\\ G(b,r)-G(a,r) & a>r_0.\end{cases}\tag{2.26}$$

[1]Theorem 2.2 still holds even when ζ is an odd integer. We omit the details here to save space.

Further, we obtain

$$\lim_{r_0/r\to 0} r^2 F(r_0, r) = r_0^{2-\zeta}, \tag{2.27}$$

$$\lim_{x/r\to 0} G_1(x, r) = \frac{2(a-1)}{(a-2)(a-3)\pi}, \tag{2.28}$$

$$\lim_{x/r\to 0} G_2(x, r) = \frac{2}{(\zeta - 2)}. \tag{2.29}$$

Then, by taking limit on both sides of (2.21), we obtain

$$\lim_{r\to\infty} d_0 = \left(\frac{2N_0(\zeta - 2) + 2\zeta r_0^{\zeta-2} \rho^* \pi \beta_K P}{\zeta r_0^{2-\zeta} N_0(1-\rho^*)(\zeta - 2)} \right)^{\frac{1}{\zeta-2}}. \tag{2.30}$$

As we can see from (2.23), the distance threshold converges to a constant when the network radius goes to infinity. This means that the number of non-zero entries per row (i.e., corresponding to each RRH) or per column (i.e., corresponding to each mobile user) in $\widehat{\mathbf{H}}$ does not scale with the network radius r in a large C-RAN. In Table 2.1, both the distance threshold d_0 and the percentages of non-zero entries in matrix $\widehat{\mathbf{H}}$ are listed for different network sizes, with $\beta_K = 10/\text{km}^2$, $\frac{P}{N_0} = 80\,\text{dB}$ and $\rho^* = 0.95$. It can be seen that, when r is large, d_0 does not change much with the network radius r. Moreover, as shown in Table 2.1, only a very low percentage entries (say 2–0.13%) in $\widehat{\mathbf{H}}$ are non-zero. That is, each RRH only needs to obtain CSI of a small number of closest users and the channel estimation overhead can be significantly reduced. If a larger SINR loss can be tolerated, the amount of CSI needed can be further reduced as shown in Table 2.2, which lists the percentages of non-zero entries in $\widehat{\mathbf{H}}$ for different ρ^*, with $\beta_K = 10/\text{km}^2$, $\frac{P}{N_0} = 80\,\text{dB}$ and $r = 10\,\text{km}$. We can see that the percentage of non-zero entries can be reduced from

Table 2.1 Percentage of non-zero entries in the channel matrix with $\beta_K = 10/\text{km}^2$, $\frac{P}{N_0} = 80\,\text{dB}$ and $\rho^* = 0.95$

r (km)	5	10	15	20
d_0 (m)	694	705	707	708
Percentage of non- zero entries (%)	1.93	0.50	0.20	0.13

Table 2.2 Percentage of non-zero entries in the channel matrix with $\beta_K = 10/\text{km}^2$, $\frac{P}{N_0} = 80\,\text{dB}$ and $r = 10\,\text{km}$

ρ^*	0.90	0.93	0.96	0.99
d_0 (m)	456	572	807	1812
Percentage of non-zero entries (%)	0.21	0.33	0.65	3.28

3.28 to 0.21% by decreasing the SINR performance from 99 to 90%, which means the sparsity of $\widehat{\mathbf{H}}$ can be increased a lot and the estimation overhead can be reduced.

2.3 Numerical Results

In this section, we first verify our analysis through numerical simulations. We then illustrate the effect of SINR ratio requirement on the choice of the distance threshold. Unless stated otherwise, we assume that the minimum distance between RRHs and users is 1 m, the path loss exponent is 3.7, and the average transmit SNR at the user side equals to 80 dB. That is $\frac{P}{N_0} = 80\,\text{dB}$.

2.3.1 *Verification*

We first verify Theorem 1 in Fig. 2.2, where $K = 1000$ and $r = 5\,\text{km}$. Figure 2.2 plots the average SINR performance ratio against the distance threshold. The SINR ratios with different N are plotted as the blue curves. Meanwhile, the lower bound $\underline{\rho}(d_0)$ derived based on the distribution in (2.19) is plotted as the red curve. It can be seen that the gap between the lower bound and the actual SINR ratios is small,

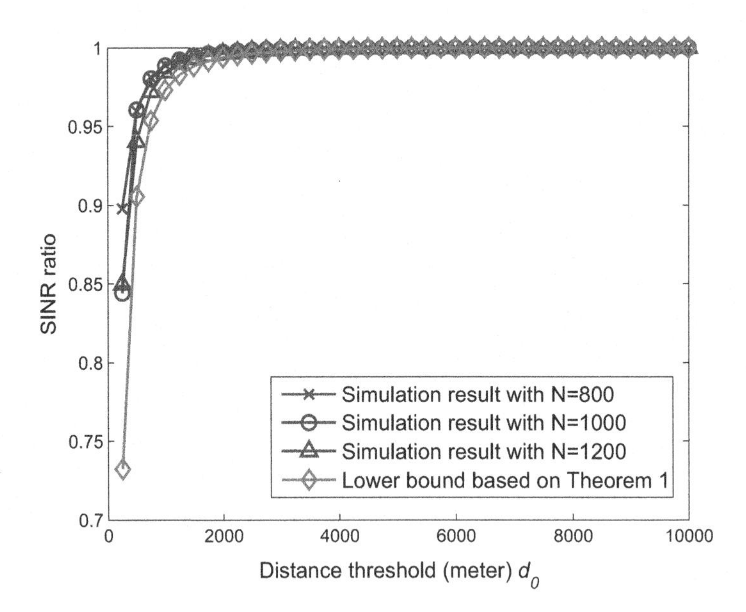

Fig. 2.2 Average SINR ratio vs distance threshold when $K = 1000$, $r = 5\,\text{km}$

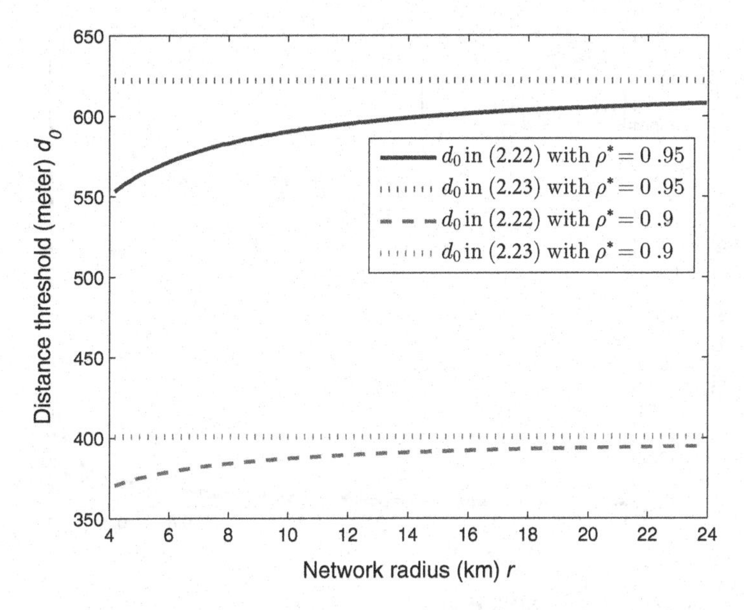

Fig. 2.3 Distance threshold d_0 vs area radius r when the user density $\beta_K = 8/\text{km}^2$

implying that the lower bound is tight. Also, it is worth noting that the lower bound derived based on Theorem 2.1 remains unchanged for different N.

We then verify in Fig. 2.3 our claims in (2.23). That is, the distance threshold converges to a constant when the network radius r becomes large. Here, the user density is $\beta_K = 8/\text{km}^2$, and the SINR ratio requirement is set to $\rho^* = 0.95$ and $\rho^* = 0.9$, respectively. As expected, the distance threshold converges quickly to a constant when the network radius increases. Indeed, the convergence is observed even when the network radius is as small as 5 km for both $\rho^* = 0.9$ and $\rho^* = 0.95$.

2.3.2 Discussion on the Distance Threshold

In Fig. 2.4, we plot the distance thresholds against the SINR ratio requirements with user density $\beta_K = 5$, 10 and $15/\text{km}^2$, respectively. The network radius is assumed to be very large. We can see that the distance threshold is very small for a wide range of ρ^*, i.e. when ρ^* is smaller than 0.95. There is a sharp increase in d_0 when ρ^* approaches to 1. This implies an interesting tradeoff: if the full SINR is to be obtained, we do need to process the full channel matrix $\mathbf{H}$ at the cost of extremely high complexity when the network size is large. On the other hand,

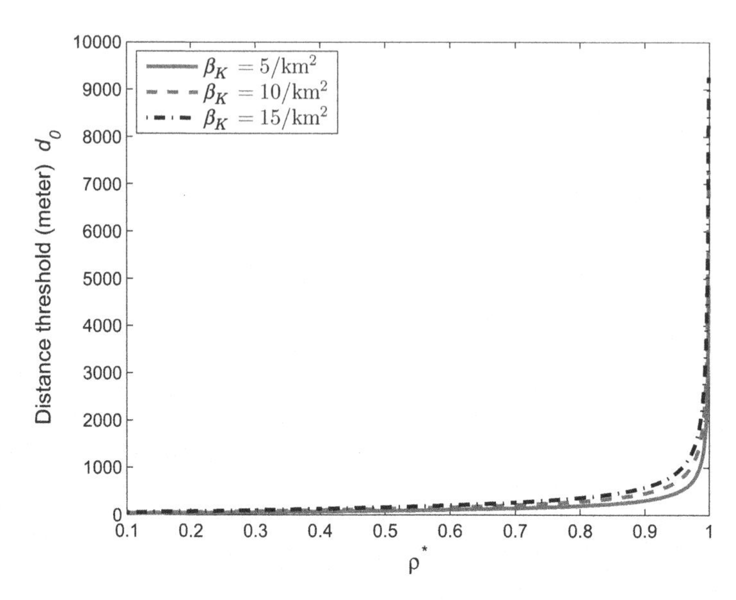

Fig. 2.4 Distance threshold vs SINR ratio

if a small percentage of SINR degradation can be tolerated, the channel matrix can be significantly sparsified, leading to low-complexity and scalable detection algorithms. We would like to emphasize that the SINR degradation may not imply a loss in the system capacity. This is because the overhead of estimating the full channel matrix can easily outweight the SINR gain. A small compromise in SINR (say reduce from 100 to 95%) may lead to a higher system capacity eventually.

2.4 Conclusions

In this chapter, we proposed a threshold-based channel matrix sparsification method, and derived a closed-form expression describing the relationship between the threshold and the SINR loss due to channel sparsification. It was shown that a vast majority of the channel coefficients can be ignored with a small percentage of SINR loss. According to our simulations, by compromising only 5% SINR loss, the CSI acquisition of each RRH can be reduced from all users to a small number of closest users. Thus, as discussed in Chap. 3, the estimation overhead can be significantly reduced. In addition, the high sparsity of channel matrix can leads to a significant reduction of computational complexity. Two low-complexity signal detection algorithms will be proposed in Chaps. 4 and 5, respectively.

References

1. C. Fan, Y. J. Zhang, and X. Yuan, "Dynamic nested clustering for parallel PHY-layer processing in cloud-RANs," *IEEE Transactions on Wireless Communications*, vol. 15, no. 3, pp. 1881–1894, 2016.
2. D. Moltchanov, "Distance distributions in random networks," *Ad Hoc Networks*, vol. 10, no. 6, pp. 1146–1166, Mar. 2012.

Chapter 3
Scalable Channel Estimation

In this chapter, we investigate the design of training sequences for time-multiplexed channel training in C-RANs. Orthogonal training sequences were shown to be optimal or nearly optimal in conventional MIMO systems, where the transmit antennas are co-located, and so are the receive antennas [3, 4]. However, orthogonal training design is very inefficient when applied to C-RAN, for that a C-RAN system usually covers a large number of users and RRHs. Allocating orthogonal training sequences to users leads to an unaffordable overhead to the system. As proved in the previous chapter, the signals from far-away users can be largely ignored without causing significant performance loss. Hence, rather than global orthogonality, we introduce the notion of *local orthogonality*, in which the training sequences of the users in the neighborhood of an RRH (i.e., the area centered around the RRH with distance below a certain threshold d_0) are required to be orthogonal to each other. The training design problem is then formulated as to find the minimum training length that preserves local orthogonality. This problem can be recast as a vertex-coloring problem, and the existing vertex-coloring algorithms [5, 6] can be applied to solve the problem. Further, we analyze the minimum training length as a function of the network size. Based on the theory of random geometric graph, we show that the training length is $O(\ln K)$ almost surely, where K is the number of users. This guarantees a scalable training-based C-RAN design, i.e., the proposed training design can be applied to a large-size C-RAN system satisfying local orthogonality at the cost of a moderate training length. Note that , on one hand, with local orthogonality, the larger the neighborhood of an RRH, the more interference from the neighboring users can be eliminated in channel estimation. Then, local orthogonality achieves a channel-estimation accuracy close to that of global orthogonality. On the other hand, a larger neighborhood area implies more channel coefficients to be estimated, thereby incurring a greater overhead to the system. As such, there is a balance to strike between the accuracy and the overhead in channel estimation. In this chapter, we study this tradeoff from the perspective of throughput maximization. We show that, with local orthogonality, the optimal d_0

Y.-J. A. Zhang et al., *Scalable Signal Processing in Cloud Radio Access Networks*, SpringerBriefs in Electrical and Computer Engineering,
https://doi.org/10.1007/978-3-030-15884-2_3

for throughput maximization can be numerically determined. The material in this chapter is mainly based on [1].

3.1 System Model

Consider a C-RAN consisting of N RRHs and K users randomly distributed over a service area. The RRHs are connected to a BBU pool by the fronthaul. We assume that the capacity of the fronthaul is unlimited, so that the signals received by the RRHs are forwarded to the BBU pool without distortion for centralized signal processing. We also assume that users and RRHs are uniformly distributed over the service area that is a square with side length r_0. The result in this chapter can be extended to service areas with other shapes. We consider a multiple-access scenario where users simultaneously transmit individual data to RRHs. The channel is assumed to be block-fading, i.e., the channel remains invariant within the coherence time of duration T.

Each transmission frame consists of T channel uses. Then, the received signal of RRH i at time t is given by

$$y_{i,t} = \sum_{k=1}^{K} d_{i,k}^{-\frac{\zeta}{2}} \gamma_{i,k} x_{k,t} + n_{i,t}, i = 1, \ldots, N, t = 1, \ldots, T \tag{3.1}$$

where $x_{k,t}$ denotes the signal transmitted by user k at time t, $n_{i,t} \sim \mathcal{CN}(0, N_0)$ is the white Gaussian noise at RRH i, $\gamma_{i,k}$ is the small-scale fading factor from user k to RRH i and is independently drawn from $\mathcal{CN}(0, 1)$, and $d_{i,k}^{-\frac{\zeta}{2}}$ represents the large-scale fading factor from user k to RRH i with $d_{i,k}$ being the distance between user k and RRH i, and ζ being the path loss exponent. Denote by $\mathbf{y}_i = [y_{i,1}, \ldots, y_{i,T}] \in \mathbb{C}^{1\times T}$ the received signal at RRH i in a vector form and $\mathbf{x}_k = [x_{k,1}, \ldots, x_{k,T}] \in \mathbb{C}^{1\times T}$ the corresponding transmitted signal vector of user k. The signal model in (3.1) can be rewritten as

$$\mathbf{y}_i = \sum_{k=1}^{K} d_{i,k}^{-\frac{\zeta}{2}} \gamma_{i,k} \mathbf{x}_k + \mathbf{n}_i, i = 1, \ldots, N \tag{3.2}$$

where $\mathbf{n}_i = [z_{i,1}, \ldots, z_{i,T}] \in \mathbb{C}^{1\times T}$ is the noise vector at RRH i. The power constraint of user k is given by

$$\frac{1}{T} \|\mathbf{x}_k\|_2^2 \leq P_k, \quad k = 1, \ldots, K \tag{3.3}$$

where P_k is the power budget of user k.

We note that if $d_{i,k} = d$ for a certain constant d for all i and k, the system in (3.2) reduces to a conventional MIMO system where both users and RRHs are co-located. If $d_{i,k} = d_k$ for all i and $d_k \neq d_{k'}$ for $k \neq k'$, then the system in (3.2) reduces to a multiuser system where the RRHs are co-located but the users are separated. In this chapter, we consider a general situation that $d_{i,k} \neq d_{i',k'}$ for $i \neq i'$ or $k \neq k'$, i.e., both users and RRHs are separated from each other.

The large-scale fading coefficients only depend on user positions and vary relatively slowly. It is usually much easier to acquire the knowledge of large-scale fading coefficients than to acquire the small-scale fading coefficients $\{\gamma_{i,k}\}$.[1] Hence, we assume that the path loss coefficients $\{d_{i,k}^{-\frac{\zeta}{2}}\}$ are known at RRHs, while $\{\gamma_{i,k}\}$ need to be estimated based on the received data in a frame-by-frame manner.

In this chapter, we aim to design an efficient transmission scheme to jointly estimate the small-scale fading coefficients $\{\gamma_{i,k}\}$ and detect the signals $\{\mathbf{x}_k\}$. Each transmission frame consists of two phases, namely, the training phase and the data transmission phase. During the training phase, users transmit training sequences to RRHs for channel estimation. During the data transmission phase, users' data are transmitted and detected at the BBU pool based on the estimated channel. More details follow.

3.1.1 *Training Phase*

Without loss of generality, let αT be the number of channel uses assigned to the training phase, where $\alpha \in (0, 1)$ is a time-splitting factor to be optimized. We refer to αT as the training length. From (3.2), the received signal at RRH i for the training phase is given by

$$\mathbf{y}_i^{\mathrm{p}} = \sum_{k=1}^{K} d_{i,k}^{-\frac{\zeta}{2}} \gamma_{i,k} \mathbf{x}_k^{\mathrm{p}} + \mathbf{n}_i^{\mathrm{p}}, i = 1, \ldots, N \tag{3.4}$$

where $\mathbf{y}_i^{\mathrm{p}} \in \mathbb{C}^{1\times\alpha T}$ is the received signal at RRH i, $\mathbf{x}_k^{\mathrm{p}} = [x_{k,1}, \ldots, x_{k,\alpha T}] \in \mathbb{C}^{1\times\alpha T}$ is the training sequence transmitted by user k, and $\mathbf{n}_i^{\mathrm{p}} \in \mathbb{C}^{1\times\alpha T}$ is the corresponding additive noise. The received signals $\{\mathbf{y}_i^{\mathrm{p}}\}$ are then forwarded to the BBU pool

[1]For example, in the LTE standard [22, 23], there are two kinds of uplink reference signals: the demodulation reference signals (DMRS) and the sounding reference signals (SRS). The DMRS is designed to estimate the small-scale fading coefficients and is transmitted every 1/2 subframe of 0.5 ms. The SRS is design to estimate the large-scale fading coefficients and is transmitted every 2 subframes at most and every 32 frame (320 subframes) at least. It is clear that the resource spending on estimating the large-scale fading coefficients is much less than that of estimating the small-scale fading coefficients. Since large-scale fading coefficients change much slowly, they can be obtained by averaging the received signal power corresponding to the SRS.

through the high-capacity fronthaul, and the channel coefficients $\{\gamma_{i,k}\}$ are estimated at the BBU pool.

The power constraint for user k in the training phase is given by

$$\frac{1}{\alpha T}\|\mathbf{x}_k^{\mathrm{p}}\|_2^2 \leq \beta_k P_k, \quad k = 1, 2, \ldots, K \tag{3.5}$$

where β_k represents the power coefficient of user k during the training phase.

3.1.2 Data Transmission Phase

In the data transmission phase, the data of users are transmitted to RRHs and then forwarded to the BBU pool through the fronthaul. The BBU pool performs coherent detection based on the estimated channel obtained in the training phase. From (3.2), the received signal of RRH i in the data transmission phase is written as

$$\mathbf{y}_i^{\mathrm{d}} = \sum_{k=1}^{K} d_{i,k}^{-\frac{\zeta}{2}} \gamma_{i,k}\mathbf{x}_k^{\mathrm{d}} + \mathbf{n}_i^{\mathrm{d}}, i = 1, \ldots, N \tag{3.6}$$

where $\mathbf{x}_k^{\mathrm{d}} = [x_{k,\alpha T+1}, \ldots, x_{k,T}] \in \mathbb{C}^{1\times(1-\alpha)T}$ is the data signal of user k, $\mathbf{y}_i^{\mathrm{d}}$ is the corresponding received signal at RRH i, and $\mathbf{n}_i^{\mathrm{d}}$ is the corresponding noise. Upon receiving the signals $\{\mathbf{y}_i^{\mathrm{d}}\}$ from users, RRHs forward $\{\mathbf{y}_i^{\mathrm{d}}\}$ to the BBU pool through the fronthaul for centralized processing. Then, the BBU pool detect the signals $\{\mathbf{x}_k^{\mathrm{d}}\}$ based on the signals $\{\mathbf{y}_i^{\mathrm{d}}\}$ and the estimated channels obtained in the training phase.

The power constraint of user k in the data transmission phase is given by

$$\frac{1}{(1-\alpha)T}\|\mathbf{x}_k^{\mathrm{d}}\|_2^2 \leq \beta_k' P_k \tag{3.7}$$

where the coefficient $\beta_k' = \frac{1-\alpha\beta_k}{1-\alpha}$ satisfies the power constraint in (3.3).

3.2 Problem Formulation

3.2.1 Throughput Optimization

The mutual information throughput is a commonly used performance measure for training-based systems [3, 4]. The throughput expression for the proposed training based scheme is derived in the Appendix. The system design problem can be formulated as to maximize the throughput over the training sequence $\{\mathbf{x}_k^{\mathrm{p}}\}$, the training length αT, the number of users K, and the power allocation coefficients

$\{\beta_k\}$ subject to the power constraints in (3.3) and (3.5). Similar problems have been previously studied in the literature. For example, when users are co-located and so are RRHs, the model in (3.2) reduces to a conventional point-to-point MIMO system. The optimal training design for throughput maximization was discussed in [3]. Specifically, the optimal strategy is to select a portion of active users while the others keep silent in transmission. The optimal number of active users is equal to $\frac{T}{2}$, and each active user is assigned with an orthogonal training sequence.[2] Moreover, when only RRHs are co-located, the model in (3.1) reduces to a multiuser MIMO system. In this case, users are randomly distributed and suffer from the near-far effect. It was shown in [4] that the optimal number of active users is in general less than $\frac{T}{2}$, but the orthogonal training design is still near-optimal. A key technique used in [3, 4] is the rotational invariance of the channel distribution when users or RRHs or both are co-located.

This chapter is focused on the training design for a general C-RAN setting where neither RRHs nor users are co-located. The rotational invariance property of the channel in general does not hold in our setting, and therefore the analysis in [3] and [4] are no longer applicable. As suggested by the optimal design in [3] and [4], it is desirable to design orthogonal training sequences for C-RAN. The challenge is that a C-RAN usually serves a large number of active users. Thus, assigning every user with an orthogonal training sequence leads to an unaffordable overhead. This inspires us to design the so-called *locally orthogonal* training sequences for C-RANs, as detailed below.

3.2.2 Local Orthogonality

The main advantage of using orthogonal training sequences is that the training signal from one user does not interfere with the training signals from other users. However, the number of available orthogonal sequences is limited by the training length αT. This quantity should be kept small so as to reduce the cost of channel estimation.

A crucial observation in a C-RAN is that both RRHs and users are randomly scattered over a large area. Thus, due to the severe propagation attenuation of electromagnetic waves over distance, the interference from far-off users can be largely ignored when processing the received signal of an RRH. This fact inspires the introduction of the channel sparsification approaches in [24–26]. These approaches were originally proposed to reduce the implementation complexity and computation complexity. In contrast, in this chapter, we use channel sparsification as a tool to identify the most interfering users in the received signal of each RRH. We only assign orthogonal training sequences to the most interfering users and ignore the rest, hence the name *local orthogonality*.

[2]For the MIMO system in [3], the total transmission power is constrained by a constant invariant to the number of users. The case that the total power linearly scales with K was discussed in [4].

We basically follow the channel sparsification approach in [26]. The only difference is that here the l_∞ norm[3] is adopted as a measure of the distance between two nodes. Specifically, the channel sparsification is to ignore relatively weak channel links based on the following criteria:

$$\tilde{\gamma}_{i,k} = \begin{cases} \gamma_{i,k}, & \|\mathbf{b}_i - \mathbf{u}_k\|_\infty < d_0 \\ 0, & \text{otherwise} \end{cases} \tag{3.8}$$

where d_0 is a predefined threshold, $\mathbf{u}_k \in \mathbb{R}^{1\times 2}$ and $\mathbf{b}_i \in \mathbb{R}^{1\times 2}$ denote the coordinates of user k and RRH i, respectively.

We now present graphical illustrations of C-RANs after channel sparsification. Denote by

$$\mathcal{U} = \{1, \cdots, K\} \tag{3.9}$$

the set of user indexes and by

$$\mathcal{B} = \{1, \cdots, N\} \tag{3.10}$$

the set of RRH indexes. Define

$$\mathcal{B}_k \triangleq \{i \mid \|\mathbf{b}_i - \mathbf{u}_k\|_\infty < d_0\}, \text{ for } k \in \mathcal{U} \tag{3.11}$$

$$\mathcal{U}_i \triangleq \{k \mid \|\mathbf{b}_i - \mathbf{u}_k\|_\infty < d_0\}, \text{ for } i \in \mathcal{B} \tag{3.12}$$

$$\mathcal{U}_i^c \triangleq \text{the complement of } \mathcal{U}_i, \text{ for } i \in \mathcal{B}. \tag{3.13}$$

The graphical representation of a C-RAN after channel sparsification is illustrated in Fig. 3.1a, where each user is connected to an RRH by an arrow if the distance between the user and the RRH is below the threshold d_0. Alternatively, the system after channel sparsification can also be represented as a bipartite graph shown in Fig. 3.1b, where each black node represents an RRH, and each white node represents a user.[4] Note that, based on the channel sparsification criteria in (3.8), a small number of users may be turned off when no RRH falls into the discs with radius d_0 centered around these users. We refer to this situation as *user outage*.

[3]For a vector $\mathbf{x} = [x_1, x_2, \ldots, x_N]$, $\|\mathbf{x}\|_\infty = \max\{|x_1|, \ldots, |x_N|\}$. The reason for adopting the l_∞ norm is to facilitate our analysis on the minimum training length in Sect. 3.4, where the random geometric graph theory developed in [31] directly applies. Other types of norms can also be used for channel sparsification. The random geometric graph theory based on a generic norm can be found, e.g., in [35].

[4]C-RAN pools BBU resources together and breaks up cell-based associations between RRHs and BBUs. That is, in theory, every user is simultaneously served by all RRHs. However, in reality, cell-based associations can still be valid in C-RAN, except that C-RAN does facilitate centralized cooperation due to centralized BBU pool and allow a single user to be served by multiple RRHs with one RRH as the leading serving RRH of the user, as suggested in Fig. 3.2. In this way, each RRH can still have a cell ID and represent a separate cell, following the convention of a cellular system.

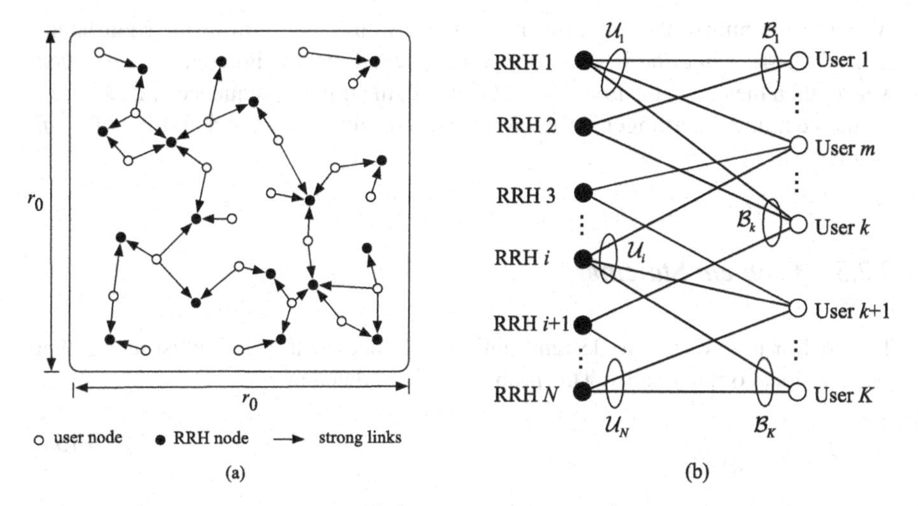

Fig. 3.1 (**a**) Graphical representation of a C-RAN after channel sparsification. (**b**) The bipartite graph representation of a C-RAN after channel sparsification

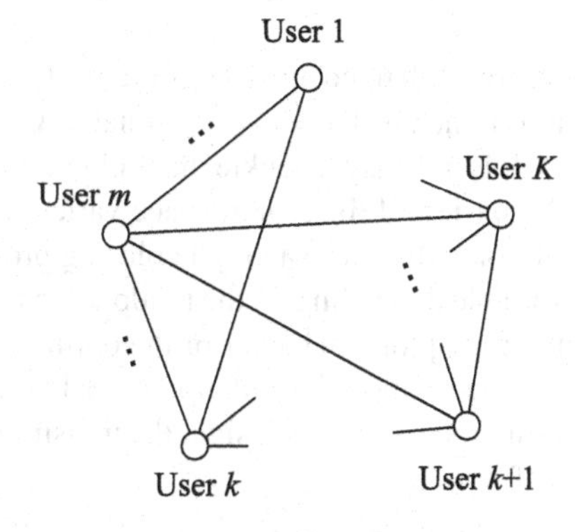

Fig. 3.2 A new graphical representation of the graph in Fig. 3.1

Numerical simulations show that the probability of user outage is close to zero with an appropriate choice of d_0. Furthermore, in practice, zero user outage can be guaranteed by slightly modifying the proposed scheme. For example, we may require that the RRHs are uniformly (rather than randomly) placed over the service area, so as to ensure that there is at least one strong RRH link for each user, no matter where it locates. More detailed discussions are, however, out of the scope of this book.

With the above channel sparsification, the received signal at RRH i in (3.4) can be rewritten as

$$\mathbf{y}_i^{\mathrm{p}} = \sum_{k \in \mathcal{U}_i} d_{i,k}^{-\frac{\xi}{2}} \gamma_{i,k} \mathbf{x}_k^{\mathrm{p}} + \sum_{k \in \mathcal{U}_i^c} d_{i,k}^{-\frac{\xi}{2}} \gamma_{i,k} \mathbf{x}_k^{\mathrm{p}} + \mathbf{n}_i^{\mathrm{p}}. \tag{3.14}$$

We aim to minimize the multiuser interference in the first term of the right-hand-side of (3.14), while the interference in the second term is ignored as it is much weaker than the one in the first term. To this end, the training sequences $\{\mathbf{x}_k^{\mathrm{p}}, k \in \mathcal{U}_i\}$ should be mutually orthogonal for any $i \in \mathcal{B}$. This gives a formal definition of *local orthogonality*.

3.2.3 Problem Statement

The goal of this work is to design training sequences with the shortest length that preserve local orthogonality. This problem is formulated as

$$\min_{\{\mathbf{x}_k^{\mathrm{p}}\}} \quad \alpha \tag{3.15a}$$

$$\text{s.t.} \quad \mathbf{x}_k^{\mathrm{p}} \perp \mathbf{x}_{k'}^{\mathrm{p}}, \forall k \neq k', k \text{ and } k' \in \mathcal{U}_i, \forall i \in \mathcal{B} \tag{3.15b}$$

$$\mathbf{x}_1^{\mathrm{p}}, \cdots, \mathbf{x}_K^{\mathrm{p}} \in \mathbb{C}^{1\times\alpha T} \tag{3.15c}$$

where α defined in Sect. 3.1.1 is the time-splitting factor for the training phase; $\mathbf{a} \perp \mathbf{b}$ means that $\mathbf{a}$ is orthogonal to $\mathbf{b}$.

It is not easy to tackle the problem in (3.15) directly, partly due to the fact that the dimension of the search space varies with α. In the following, we will solve (3.15) by converting it to a graph coloring problem. In addition, the optimal α is a random variable depending on the random locations of RRHs and users. We will characterize the asymptotic behavior of the optimal α as the network size goes to infinity.

The problem formulation in (3.15) is for uplink channel estimation. With locally orthogonal training design, the transmission protocol for uplink C-RAN is presented as follows.

- Step 1: The BBU pool performs channel sparsification based on the criteria in (3.8).
- Step 2: The BBU pool computes the training sequences based on local orthogonality.
- Step 3: The training sequences are assigned to users through the control channels.
- Step 4: Each user transmits the corresponding training sequence in the training phase. Upon receiving the signals from all RRHs, the BBU pool estimates the channel.
- Step 5: Each user transmits data in the data transmission phase, and the BBU pool detects user data based on the received signal and the estimated channel.

We emphasize that the training design for the uplink directly carries over the downlink by swapping the roles of users and RRHs. That is, in the uplink phase, the training sequences are transmitted by users and the local orthogonality is preserved at the RRH side, while in the downlink, the training sequences are transmitted by RRHs and the local orthogonality is preserved at the user side. Note that, in the downlink phase, the training sequences are still computed by the BBU pool based on local orthogonality, under the assumption that the large-scale channel coefficients are available to the BBU pool. Then, the training sequences are assigned to RRHs through the fronthaul links. As such, we henceforth focus on the training design for the uplink. The proposed design straightforwardly carries over to the downlink.

3.2.4 Related Work

In the considered training-based C-RAN, orthogonal training sequences do not interfere with each other; the interference only comes from far-off users with non-orthogonal training sequences. This is similar to the problem of *pilot contamination* in multi-cell massive MIMO systems [7], where the orthogonal training sequences used in each cell are reused among cells.

There are several existing approaches to tackle the pilot contamination problem [7–21]. For example, data-aided channel estimation with superimposed training design was proposed in [9–11] to reduce the system overhead spent on channel estimation and to suppress pilot contamination. In [12, 13], the authors proposed blind channel estimation based on singular value decomposition (SVD). However, both superimposed training and blind channel estimation involve high computational complexity in implementation, especially when applied to a C-RAN with a large network size.

Time-multiplexed training design has also been considered to address the problem of pilot contamination [14–17]. The key issue is the design of the training-sequence reuse pattern among cells. In [14] and [15], users within each cell are classified into cell-edge and cell-center users. The pilots for cell-center users are reused across all cells, while orthogonal pilot sub-sets are assigned to the cell-edge users. In [16], the cells surrounding the home cell by one or more rings are assigned orthogonal pilot sets. In [17], training sequences are assigned to users based on the angle of arrival. However, the notion of cell is no longer adopted in C-RAN, as RRHs in a C-RAN are fully connected to enable full-scale cooperation. Therefore, the cell-based techniques in [14–17] are not applicable to C-RAN.

It is also worth mentioning that training-based C-RAN has been previously studied in the literature [18, 19]. In [18], the authors proposed a coded pilot design where RRHs can be turned off to avoid pilot collisions at the cost of certain performance degradation. In [19], only a portion of users are allowed to transmit pilots for channel training in a transmission block, and the channels of the other users are not updated. The performance of this scheme degrades severely when there are a large number of active users in the system, such as the case in a C-RAN. Therefore, training design for C-RAN deserves further endeavor, which is the main focus of this work.

3.3 Training Sequence Design

In this section, we solve problem (3.15) based on graph coloring. We first formulate a graph coloring problem that is equivalent to problem (3.15).

In Fig. 3.2, we define a new graph $G = \{\mathcal{U}, E\}$ with vertex set $\mathcal{U}$ and edge set E, where two users k and m in $\mathcal{U}$ are connected by an edge $e_{k,m} \in E$ if and only if there exists an RRH i with both distances $\|\mathbf{b}_i - \mathbf{u}_k\|_\infty$ and $\|\mathbf{b}_i - \mathbf{u}_m\|_\infty$ smaller than d_0. Then, the edge set E can be represented as $E = \{e_{k,m} | \mathcal{B}_k \cap \mathcal{B}_m \neq \emptyset, \forall k \neq m, k, m \in \mathcal{U}\}$. Denote by $c : \mathcal{U} \to \mathcal{C}$ a map from each user $k \in \mathcal{U}$ to a color $c(k) \in \mathcal{C}$. We then formulate the following vertex coloring problem over G:

$$\min_{c} \quad |\mathcal{C}| \tag{3.16a}$$

$$\text{s.t.} \quad c(k) \neq c(m), \text{ if } \mathcal{B}_k \cap \mathcal{B}_m \neq \emptyset, \forall k \neq m, k, m \in \mathcal{U}. \tag{3.16b}$$

Note that the solution to (3.16), denoted by $\chi(G)$, is referred to as the chromatic number of the graph in Fig. 3.2. We further have the following result.

Theorem 3.1 *The vertex coloring problem over G in* (3.16) *is equivalent to the training design problem in* (3.15).

Proof Each color can be seen as an orthogonal training sequence. Then, the color set $\mathcal{C}$ can be mapped into a set of orthogonal training sequences $\{\mathbf{x}_k^{\mathrm{p}}\}$. The cardinality of $\mathcal{C}$ equals to the number of orthogonal training sequences, i.e., $|\mathcal{C}| = \alpha T$. From (3.16b), the statement that any two vertices connected by an edge are colored differently is equivalent to the statement that any two users with distances to a common RRH smaller than d_0 transmit orthogonal training sequences. Then, as users in $\mathcal{U}_i$ all have distances smaller than d_0 to RRH i, any two users in $\mathcal{U}_i$ must be connected by an edge in the new graph G. This is equivalent to say that the training sequences assigned to users in $\mathcal{U}_i$ are orthogonal to each other. Therefore, (3.16b) is equivalent to (3.15b), which concludes the proof.

We now discuss solving the vertex coloring problem in (3.16). This is a well-known NP-complete problem [27, 28]. Exact solutions can be found, e.g., using the algorithms proposed in [29, 30]. However the running time of these algorithms is acceptable only when the corresponding graph has a relatively small size. For a large-size graph as in a C-RAN, algorithms in [5, 6] are preferable to yield suboptimal solutions with much lower complexity. In this chapter, we adopt in the simulations the *DSATUR algorithm*, which is a greedy-based low-complexity algorithm with near optimal performance [6]. The DSATUR algorithm mainly consists of two steps. The first step is to find the vertex with the maximum saturation degree.[5] In the second step, we use an existing color to color the chosen vertex. If

[5]For a vertex u, the (ordinary) degree is defined as the number of vertices connected to u; the saturation degree is defined as the number of vertices in distinct colors connected to u.

none of the existing colors is feasible, we assign a new color to the chosen vertex. The above two steps are executed iteratively until all the vertices are colored. For completeness, we present the details of the *DSATUR algorithm* in Algorithm 3.1. Note that the DSATUR algorithm is a greedy-based algorithm and colors the vertices in a one-by-one manner. Thus, the convergence of the algorithm is naturally ensured.

Algorithm 3.1 The DSATUR Algorithm for Graph Coloring

1: Initial $\mathcal{A} = \{k = 1, \ldots, K\}$ and $\mathcal{C} = \emptyset$.
2: Set $t \Leftarrow t + 1$
3: **While** $\mathcal{A} \neq \emptyset$
  Step 1: Select $k \in \mathcal{A}$ with maximum saturation degree.
    If two vertexes have the same maximum saturation degree, choose the one with maximum ordinary degree.
  Step 2: Color k greedily so that
    $c(k) = \min\{i \in \mathcal{C} | c(m) \neq i, \forall m \in \{m | e_{k,m} \in E\}\}$
    if $c(k) = \emptyset$
      $c(k) = |\mathcal{C}| + 1$
      $\mathcal{C} = \mathcal{C} \cup \{c(k)\}$
    end if
  Step 3: $\mathcal{A} = \mathcal{A} \backslash k$
4: **end while**

We now discuss the construction of training sequences $\{\mathbf{x}_k^{\mathrm{p}}\}$ for problem (3.15) based on the coloring pattern c and the chromatic number $\chi(G)$ obtained from solving (3.16). We first generate $\chi(G)$ orthonormal training sequences of length $\chi(G)$, i.e. $\tilde{\mathbf{x}}_i^{\mathrm{p}}(\tilde{\mathbf{x}}_i^{\mathrm{p}})^{\mathrm{H}} = 1, \forall i$, and $\tilde{\mathbf{x}}_i^{\mathrm{p}}(\tilde{\mathbf{x}}_j^{\mathrm{p}})^{\mathrm{H}} = 0, \forall i, j \in \{1, \ldots, \chi(G)\}$ with $i \neq j$. Then, the training sequence $\mathbf{x}_k^{\mathrm{p}}$ for user k is scaled to meet the power constraint in (3.5), i.e. $\mathbf{x}_k^{\mathrm{p}} = \sqrt{\chi(G)\beta_k P_k} \tilde{\mathbf{x}}_{c(k)}^{\mathrm{p}}$, $k \in \mathcal{U}$.

3.4 Optimal Training Length

In the preceding section, we proposed a training design algorithm for the problem in (3.15), and obtained that the minimum training length αT is given by the chromatic number $\chi(G)$. In this section, we focus on the behavior of the training length as the network size increases. We show that the training length scales at most at the rate of $O(\ln K)$ as the network size goes to infinity under the assumption of a fixed user density.

3.4.1 Graph with Infinite RRHs

Our analysis is based on the theory of random geometric graph. Recall from (3.8) that the edge generation of graph G in Fig. 3.2 follows the rule below:

$$E = \{e_{k,m} | \exists i, \|\mathbf{u}_k - \mathbf{b}_i\|_\infty < d_0 \text{ and } \|\mathbf{u}_m - \mathbf{b}_i\|_\infty < d_0\}. \tag{3.17}$$

This is unfortunately different from the edge-generation rule of a random geometric graph [31]. To circumvent this obstacle, we introduce a new graph as follows:

$$G^\infty = \{\mathcal{U}, E^\infty\} \text{ with } E^\infty = \{e_{k,m}^\infty | \|\mathbf{u}_k - \mathbf{u}_m\|_\infty < 2d_0\}. \tag{3.18}$$

We have the following result.

Lemma 3.1 *The graph G is a subgraph of G^∞.*

Proof Since G and G^∞ have a common vertex set $\mathcal{U}$, we only need to prove $E \subseteq E^\infty$. For any $e_{k,m} \in E$, we have $\|\mathbf{u}_k - \mathbf{u}_m\|_\infty = \|\mathbf{u}_k - \mathbf{b}_i + \mathbf{b}_i - \mathbf{u}_m\|_\infty \leq \|\mathbf{u}_k - \mathbf{b}_i\|_\infty + \|\mathbf{u}_m - \mathbf{b}_i\|_\infty < 2d_0$. Hence, $e_{k,m} \in E^\infty$.

Graph G^∞ can be seen as a supergraph[6] of G with infinite RRHs, i.e., there always exists an RRH located between two users provided that the distance between these two users does not exceed $2r$. Since G is a subgraph of G^∞, the chromatic number $\chi(G^\infty)$ serves as an upper bound for $\chi(G)$.

3.4.2 Asymptotic Behavior of the Training Length

We are now ready to characterize the asymptotic behavior of the training length as $K \to \infty$. Denote by $\delta = \frac{K}{r_0^2}$ the user density in the service area. We have the following theorem.

Theorem 3.2 *As $K \to \infty$ and $\frac{\delta d_0^2}{\ln K} \to \rho$, the minimum training length αT preserving local orthogonality satisfies*

$$\limsup_{K\to\infty} \left(\frac{\alpha T}{\delta d_0^2}\right) \leq 4 f^{-1}\left(\frac{1}{4\rho}\right), \; a.s., \tag{3.19}$$

where $\rho \in (0, +\infty)$ is a predetermined constant, and f^{-1} is the inverse function of $f(x)$ over the domain $[1, +\infty)$ with $f(x)$ defined as below

[6] A supergraph is a graph formed by adding vertices, edges, or both to a given graph. If H is a subgraph of G, then G is a supergraph of H.

$$f(x) = 1 - x + x \ln x, \quad x > 1 \tag{3.20}$$

and a.s. stands for almost surely.

Proof From Lemma 3.1, G is a subgraph of G^∞, and $\chi(G)$ is upper bounded by $\chi(G^\infty)$. As the minimum αT is given by $\chi(G)$, it suffices to show

$$\limsup_{K\to\infty} \left(\frac{\chi(G^\infty)}{\delta d_0^2} \right) \le 4 f^{-1}\left(\frac{1}{4\rho}\right), \; a.s. \tag{3.21}$$

Denote by $\omega(G^\infty)$ the clique number[7] of G^∞. From Theorem 2 in [31], we have[8]

$$\limsup_{K\to\infty} \left(\frac{\omega(G_v^\infty)}{\delta d_0^2} \right) \le 4 f^{-1}\left(\frac{1}{4\rho}\right), \; a.s. \tag{3.22}$$

The clique number $\omega(G^\infty)$ serves as a lower bound for $\chi(G^\infty)$ in general. However, as $K \to \infty$, the clique number almost surely converges to the chromatic number [32], i.e.

$$\lim_{K\to\infty} \frac{\chi(G^\infty)}{\omega(G^\infty)} = 1, \; a.s. \tag{3.23}$$

Combining (3.22) and (3.23), we obtain

$$\limsup_{K\to\infty} \left(\frac{\chi(G^\infty)}{\delta d_0^2} \right) \le 4 f^{-1}\left(\frac{1}{4\rho}\right), \; a.s. \tag{3.24}$$

This completes the proof.

Remark 3.1 In Theorem 3.2, the meaning of the constant ρ is explained as follows. Recall that δ is the user density. Thus, δd_0^2 is the average number of users with distance smaller than d_0 to an RRH. Therefore, $\frac{\delta d_0^2}{\ln K} \to \rho$ implies that the average number of users with distance smaller than d_0 to an RRH scales at the speed of $\rho \ln K$ as $K \to \infty$.

Remark 3.2 We note that the upper bound of the minimum training length (αT) in Theorem 3.2 is independent of the number of RRHs (N). For a finite N, the minimum training length increases with N, and is upper bounded by $\chi(G^\infty)$.

[7] A complete graph is a graph where every pair of distinct vertices are connected by a unique edge. The clique number is defined as the number of vertices of the largest complete subgraph of G^∞.

[8] To invoke Theorem 2 in [31], the distance threshold in [31] is set to $\frac{2r}{r_0}$.

Remark 3.3 From Theorem 3.2, the training length αT scales as $4\rho f^{-1}\left(\frac{1}{4\rho}\right)\ln K$. Note that $4\rho f^{-1}\left(\frac{1}{4\rho}\right)$ monotonically increases with ρ. Therefore, a larger ρ results in a longer training length. This is reasonable since that, from Remark 3.1, a larger ρ means on average a larger amount of users are surrounding an RRH, and thus more training sequences are required to preserve the local orthogonality.

From Theorem 3.2, we also claim the following result.

Corollary 3.1 *As $K \to \infty$ and $r_0^2 \to \infty$ with $\delta = \frac{K}{r_0^2}$ and d_0 fixed, the minimum training length to preserve local orthogonality scales at most in the order of* $\ln K$.

Proof To meet the constraint in Theorem 3.2 that $\frac{\delta d_0^2}{\ln K} \to \rho$, one way is to fix d_0^2 and let δ scale in the order of $\ln K$. Note that a graph with a fixed user density δ can be generated by randomly deleting users from a graph with user density $\delta = O(\ln K)$. This implies that the minimum training length for the case of δ fixed is upper bounded by that of the case of $\delta = O(\ln K)$. From Theorem 3.2, when $\delta = O(\ln K)$, the minimum training length scales in the order of $\ln K$. Therefore, the minimum training length for a fixed δ scales at most in the order of $\ln K$, which completes the proof.

Remark 3.4 Corollary 3.1 indicates that the minimum training length of our proposed training design scheme is moderate even for a large-scale C-RAN.

3.4.3 Further Discussions

In the above discussions, we assume that the chromatic number of G can be determined accurately. However, the algorithms proposed in [29, 30] to find the chromatic number always have high computation complexity and are not suitable for practical systems. As mentioned in Sect. 3.3, a greedy algorithm called the *DSATUR algorithm* is applied in this chapter, which cannot be guaranteed to achieve the chromatic number but with relatively low complexity. However, the problem arises whether the number of colors used by a suboptimal coloring algorithm still scales in the order of $\ln K$. For this problem, we have the following theorem.

Theorem 3.3 *The training length αT determined by the DSATUR algorithm in Algorithm 3.1 scales at most in the order of* $\ln K$ *as $K \to \infty$ and $\frac{\delta d_0^2}{\ln K} \to \rho$.*

Proof Revisit the DSATUR algorithm in Algorithm 3.1. In each coloring step, the vertex to be colored prefers to use an existing color. Then, the number of colors in every step is upper bounded by $\Delta(G)+1$, where $\Delta(G)$ is the maximum degree of G. Therefore, the number of colors used by the DSATUR algorithm is upper bounded

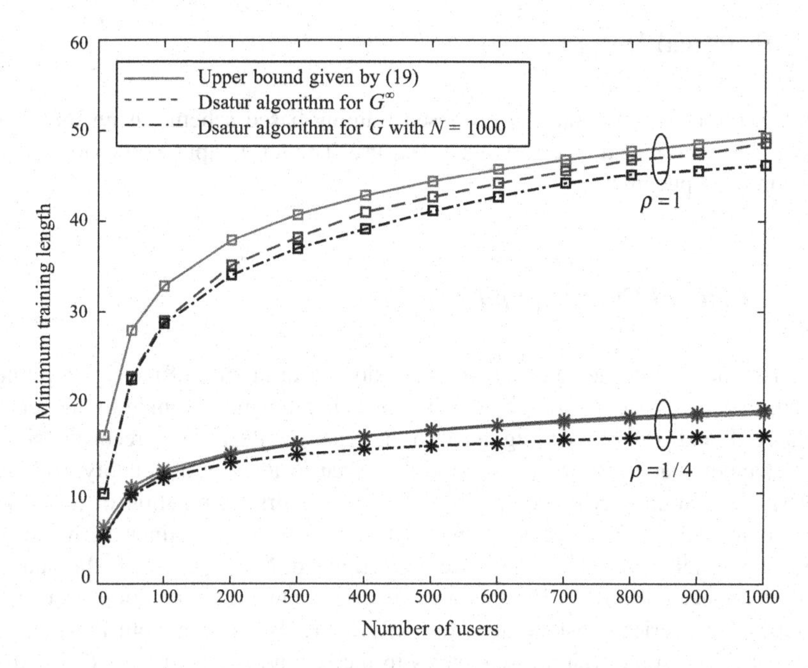

Fig. 3.3 The minimum training length against the number of users K with $N = 1000$

by $\Delta(G) + 1$. We also see from Lemma 3.1 that $\Delta(G) \leq \Delta(G^{\infty})$. Therefore, it suffices to characterize the behavior of $\Delta(G^{\infty})$. From Theorem 1 in [31], we have

$$\limsup_{K \to \infty} \left(\frac{\Delta(G^{\infty})}{\delta d_0^2} \right) \leq 16 f^{-1} \left(\frac{1}{16\rho} \right), \quad a.s. \tag{3.25}$$

where f^{-1} is given in (3.20). Together with $\delta d_0^2 = O(\ln K)$, we see that $\Delta(G^{\infty})$ scales in the order of $\ln K$. Thus, $\Delta(G)$ and the number of colors given by the DSATUR algorithm scales at most in the order of $\ln K$, which completes the proof.

Figure 3.3 gives the numerical results to verify our analysis. Both the chromatic number of G^{∞} and G with $N = 1000$ are included for comparison. In simulation, the chromatic number is found by the DSATUR algorithm and averaged over 1000 random realizations. From Fig. 3.3, the training length αT is strictly upper bounded by the theoretical result given in (3.19). We also note that the output of the DSATUR algorithm for G^{∞} may be slightly greater than the upper bound given in (3.19). This is due to the use of the suboptimal coloring algorithm in simulation. From Fig. 3.3, we see that the simulated minimum training length is close to the upper bound given in (3.19).

3.5 Practical Design

In this section, we evaluate the proposed training-based scheme using the information throughput as the performance measure. The throughput expression can be found in the Appendix.

3.5.1 Refined Channel Sparsification

We first show that the channel sparsification criteria in (3.8) can be refined to improve the system throughput while keeping the minimum training length unchanged and still preserving local orthogonality. Recall the received signal after channel sparsification shown in (3.14). Due to local orthogonality, for each RRH i, the training sequences $\{\mathbf{x}_k^{\mathrm{p}}, k \in \mathcal{U}_i\}$ in the first summation of (3.14) are orthogonal and the channel coefficients $\{\gamma_{i,k}, k \in \mathcal{U}_i\}$ are estimated, while the channel coefficients $\{\gamma_{i,k}, k \in \mathcal{U}_i^c\}$ are not estimated. Note that $|\mathcal{U}_i|$ is the number of users connected to RRH i in the bipartite graph. Due to the randomness of user locations, $|\mathcal{U}_i|$ varies considerably from RRH to RRH, as seen from Fig. 3.4. For a certain RRH, some training sequences in the second summation of (3.14) may be orthogonal to the training sequences $\{\mathbf{x}_k^{\mathrm{p}}, k \in \mathcal{U}_i\}$ in the first summation. The corresponding channel coefficients $\{\gamma_{i,k}, k \in \mathcal{U}_i^c\}$ can be estimated at RRH i without incurring additional training overhead. Note that estimating a channel coefficient $\gamma_{i,k}$ with $k \notin \mathcal{U}_i$ is equivalent to moving k from $\mathcal{U}_i^c$ to $\mathcal{U}_i$, or alternatively speaking,

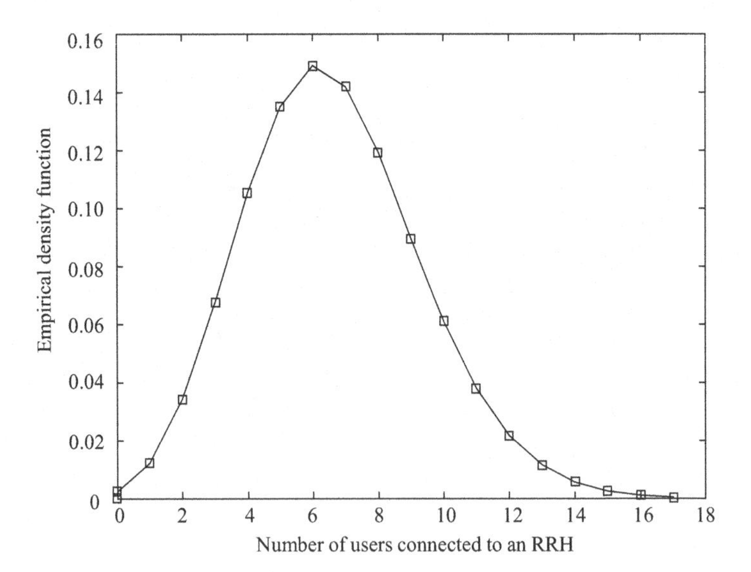

Fig. 3.4 An empirical density function of the number of users served an RRH with $N = 1000$ and $K = 1000$. Every simulated point is averaged over 1000 channel samples

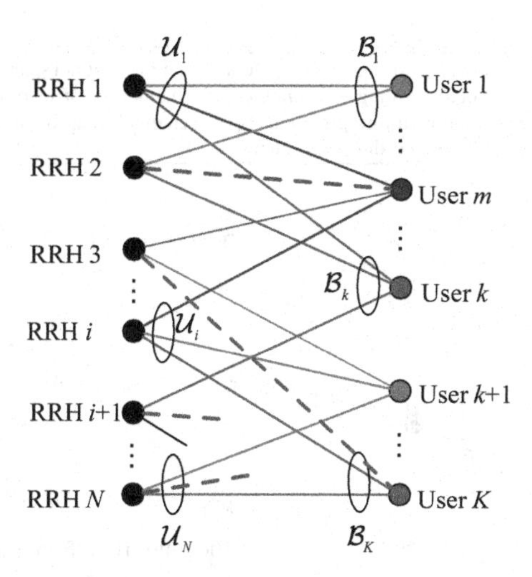

Fig. 3.5 An example to illustrate the refined channel sparsification. We use green, blue, and red colors to represent three orthogonal training sequences. The solid lines represent the strong links survived after channel sparsification. To preserve local orthogonality, users connected to a common RRH should use different colors. The dash lines denote the new edges added to RRHs without compromising the property of local orthogonality. As no new color is added to the graph, the refined channel sparsification does not increase the training overhead (Color figure online)

to adding the edge connecting user k and RRH i to the bipartite graph. As the terms in the second summation of (3.14) are treated as interference in channel estimation, reducing the size of $\mathcal{U}_i^c$ is beneficial in suppressing the interference. Therefore, it is desirable to add more edges to the bipartite graph without compromising local orthogonality.

We now show how to add more edges to each RRH. Denote by c and $\chi(G)$ the coloring pattern and the chromatic number obtained by solving (3.16), respectively. Based on c, the user set $\mathcal{U}$ can be partitioned into $\chi(G)$ subsets each with a different color. Then, each RRH i chooses the closest users colored differently from the users in $\mathcal{U}_i$. An example to this procedure is illustrated in Fig. 3.5, where solid lines represent the strong links after channel sparsification following the criteria in (3.8), and the dash lines denote the new edges added to RRHs. The coloring pattern c is obtained by solving problem (3.16).

Clearly, the refined channel sparsification in Fig. 3.5 still preserves local orthogonality. Hence, the analytical results in Sect. 3.4 are still applicable to the refined channel sparsification and the minimum training length still follows $O(\ln K)$.

3.5.2 Numerical Results

Numerical results are presented to demonstrate the performance of our proposed scheme. Users and RRHs are uniformly scattered in a square area with $r_0 = 100$ m, i.e., RRHs and users are uniformly distributed over a $100\,\text{m} \times 100\,\text{m}$ region. The

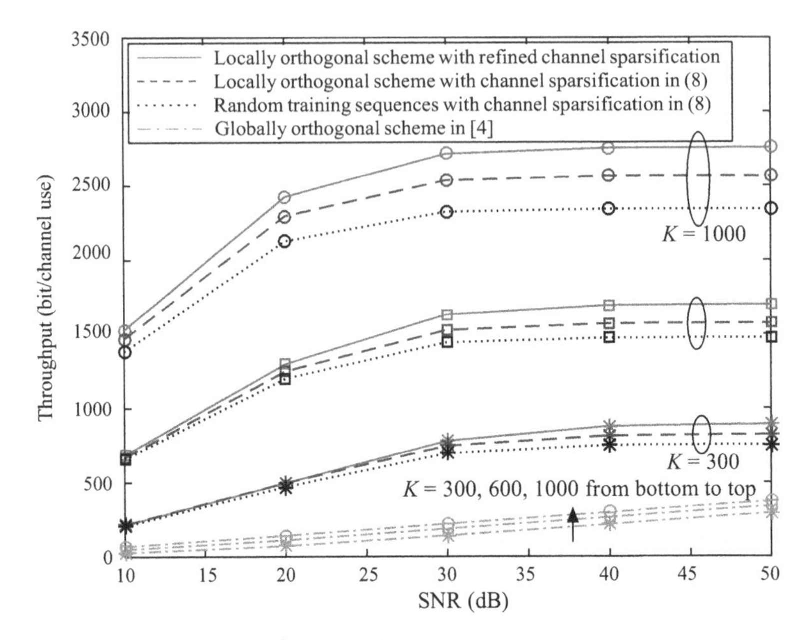

Fig. 3.6 Performance comparison between the new and the old channel sparsification criteria with $\rho = 0.5$ and $N = K = 1000, 600, 300$. SNR = P_0/N_0 and $r_0 = 100$ m

pathloss exponent is $\zeta = 3.5$. The channel coherence time is set at $T = 100$. In simulation, we assume equal power allocation among users and also between the training phase and the data transmission phase, i.e., $P_1 = P_2 = \cdots = P_K = P_0$ and $\beta_1 = \beta_2 = \cdots = \beta_K = 1$.

Performance comparison among different schemes is given in Fig. 3.6 with $\rho = 0.5$ (where ρ is defined in Theorem 3.2) and $N = K = 1000, 600, 300$ marked on the curves. From Fig. 3.6, we see that the refined channel sparsification method outperforms the channel sparsification in (3.8) by about 10%. We first compare our scheme with the globally orthogonal training design. It has been shown in [2, 3] that, for a training-based multiuser MIMO, the optimal training design for throughput maximization in the high SNR regime is to assign orthogonal training sequence to a selected set of K' users while keeping the other users silent, where K' is bounded by the coherence time T. This training design is simple but questionable for C-RAN. First, asymptotic optimality at high SNR does not guarantee good performance in the practical SNR region. In fact, as shown in Fig. 3.6, the orthogonal training design is significantly outperformed by our proposed scheme in the considered SNR range. Second, globally orthogonal training only allows a small portion of users to be active. This will cause a serious fairness problem for user access to C-RAN, especially when the number of users K in C-RAN is large.

We now compare our scheme with the random training design, where each user is assigned with a randomly generated training sequence of a certain predetermined length αT. Note that αT can be determined using the DSATUR algorithm (which

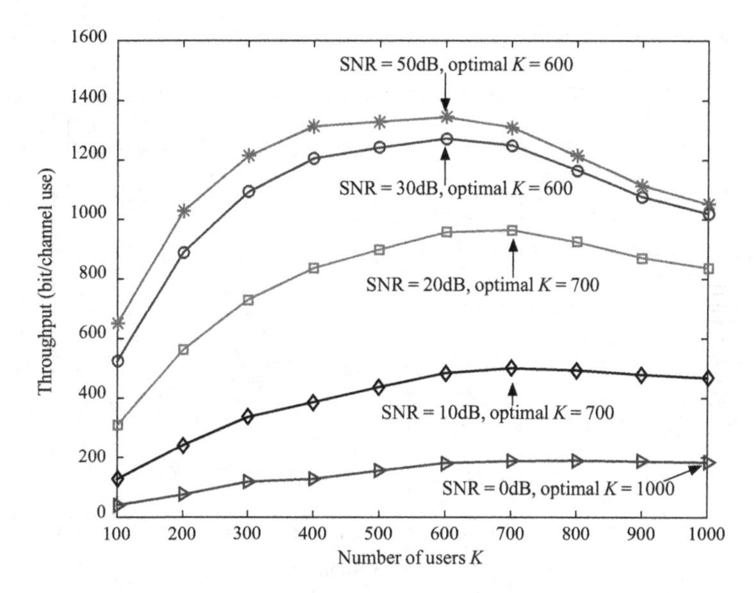

Fig. 3.7 The system throughput versus the number of users. The number of RRHs is $N = 500$. The distance threshold for channel sparsification is set to $r = 10\,\text{m}$ and the side length $r_0 = 100\,\text{m}$

ensures local orthogonality). From Fig. 3.6, we see that our proposed scheme has over 20% performance enhancement in throughput over the random training scheme in the SNR range of 30–50 dB.

The system throughput against the number of users is given in Fig. 3.7. We see that, for given N, d_0, and SNR $= P_0/N_0$, there is a tradeoff in maximizing the system throughput over K. On one hand, the total received power at each RRH increases with K, and so is the information rate for the data transmission phase. On the other hand, a larger K implies more channel coefficients to be estimated, and hence increases the required training length. Therefore, a balance needs to be stroke in optimizing K. For example, for SNR $= 30\,\text{dB}$, the optimal K occurs at $K = 600$.

A larger K implies higher received signal power at each RRH and at the same time, higher interference power at each RRH (as by channel sparsification the signals from far-off users are treated as interference). In the low SNR region, the interference is overwhelmed by the noise. Therefore, the more users, the higher system throughput, as seen from the curve for SNR $= 0\,\text{dB}$ in Fig. 3.7. In the high SNR region, the interference dominates the noise. Then, a larger K does not necessarily translate to a higher throughput.

Figure 3.8 illustrates the system throughput against the distance threshold d_0 with SNR $= 50\,\text{dB}$. We see that there is an optimal distance threshold to maximize the throughput for fixed N and K. The reason is explained as follows. With a high

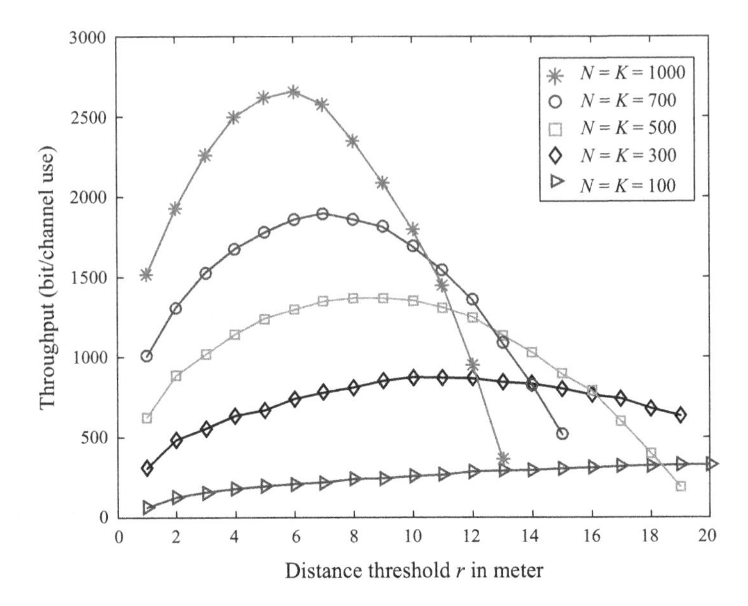

Fig. 3.8 The system throughput versus the distance threshold d_0. The signal to noise ratio SNR = $P_0/N_0 = 50\,\text{dB}$, and the side length $r_0 = 100\,\text{m}$

threshold d_0, more interference from neighboring users are eliminated in channel estimation, and thus better throughput is achieved. However, at the same time, increasing d_0 implies increasing the training length, as more training sequences are required to be orthogonal at each RRH. Therefore, there is an optimal d_0 for throughput maximization, as illustrated in Fig. 3.8.

Figure 3.9 shows the system throughput against the number of RRHs at various SNR values. From Fig. 3.9, we see that the system throughput monotonically increases with N for all the considered SNR values, implying that the more RRHs, the higher the cooperation gain. This gain is obtained by assuming full cooperation among the RRHs, at the cost of a fronthaul overhead that scales with the number of RRHs.[9]

[9]We elaborate on the overhead of full RRH cooperation as follows. For uplink transmission (as considered in Fig. 3.9), the more RRHs, the more data need to be transmitted from the RRHs to the BBU pool for centralized processing. For downlink transmission, besides the overhead for fronthaul data transmission, the cost also includes more complicated precoding at RRHs.

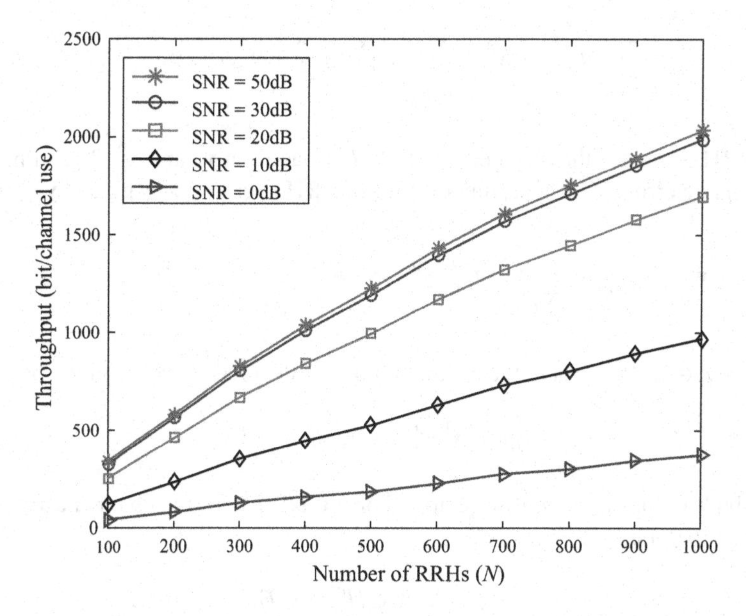

Fig. 3.9 The system throughput versus the number of RRHs. The number of users is $K = 500$. The distance threshold for channel sparsification is set to $r = 10\,\text{m}$ and the side length $r_0 = 100\,\text{m}$.

3.6 Conclusions

In this chapter, we considered training-based channel estimation for C-RANs with N RRHs and K users. We introduced the notion of local orthogonality and formulated the training design problem so as to find the minimum length of training sequences that preserve local orthogonality. A training design scheme based on graph coloring was also proposed. We further showed that the training length is $O(\ln K)$ almost surely as $K \to \infty$. Therefore, the proposed training design can be applied to a large-size C-RAN satisfying local orthogonality at the cost of an acceptable training length. This chapter was focused on minimizing the training length under the local orthogonality constraint. The joint optimization of the training sequences and the data transmission scheme to maximize the system throughput will be an interesting topic for future research.

Appendix

In this appendix, we derive a throughput lower bound for the training-based CRAN scheme described in Sect. 3.1. With channel sparsification, the received signal at RRH i in (3.4) can be rewritten as

$$\mathbf{y}_i^{\mathrm{p}} = \sum_{k \in \mathcal{U}_i} d_{i,k}^{-\frac{\zeta}{2}} \gamma_{i,k} \mathbf{x}_k^{\mathrm{p}} + \sum_{k \in \mathcal{U}_i^c} d_{i,k}^{-\frac{\zeta}{2}} \gamma_{i,k} \mathbf{x}_k^{\mathrm{p}} + \mathbf{n}_i^{\mathrm{p}}, i \in \mathcal{B}. \tag{3.26}$$

Each RRH estimates the channel $\{\gamma_{i,k}, k \in \mathcal{U}_i\}$ based on $\mathbf{y}_i^{\mathrm{p}}$ and $\mathbf{x}_k^{\mathrm{p}}$. The minimum mean-square error (MMSE) estimator [33] of RRH i for user k is given by

$$\mathbf{w}_{i,k} = d_{i,k}^{-\frac{\zeta}{2}} \mathbf{x}_k^{\mathrm{p}} \Big(\sum_{k' \in \mathcal{U}_i} d_{i,k'}^{-\zeta} (\mathbf{x}_{k'}^{\mathrm{p}})^{\mathrm{H}} \mathbf{x}_{k'}^{\mathrm{p}} + N_0 \mathbf{I} \Big)^{-1}, \ k \in \mathcal{U}_i \tag{3.27}$$

where $\mathbf{w}_{i,k} \in \mathbb{C}^{1 \times \alpha T}$. Then, the estimation of $\gamma_{i,k}$ denoted by $\hat{\gamma}_{i,k}$ is given by

$$\hat{\gamma}_{i,k} = \mathbf{y}_i^{\mathrm{p}} \mathbf{w}_{i,k}^{\mathrm{H}}, \ k \in \mathcal{U}_i, \ i \in \mathcal{B} \tag{3.28}$$

where both $\mathbf{y}_i^{\mathrm{p}}$ and $\mathbf{w}_{i,k}$ are row vectors. For $k \in \mathcal{U}_i^c$, $i \in \mathcal{B}$, the channel estimate of $h_{i,k}$ is set to 0, i.e.,

$$\hat{\gamma}_{i,k} = 0, \ k \in \mathcal{U}_i^c, \ i \in \mathcal{B}. \tag{3.29}$$

Denote by $\mathrm{MSE}_{i,k}$ the corresponding mean square error (MSE) of RRH i for user k. Then

$$\begin{aligned} \mathrm{MSE}_{i,k} &= \mathbb{E}\Big[|\gamma_{i,k} - \hat{\gamma}_{i,k}|^2 \Big] \\ &= \mathbf{w}_{i,k} \Big(\sum_{k' \in \mathcal{U}_i^c} d_{i,k'}^{-\zeta} (\mathbf{x}_{k'}^{\mathrm{p}})^{\mathrm{H}} \mathbf{x}_{k'}^{\mathrm{p}} \Big) \mathbf{w}_{i,k}^{\mathrm{H}}, \\ &\quad + 1 - d_{i,k}^{-\frac{\zeta}{2}} \mathbf{x}_k^{\mathrm{p}} \mathbf{w}_{i,k}^{\mathrm{H}} \ \text{for } k \in \mathcal{U}_i, i \in \mathcal{B}; \qquad (3.30\mathrm{a}) \\ \mathrm{MSE}_{i,k} &= 1, \quad \text{for } k \in \mathcal{U}_i^c, i \in \mathcal{B}. \qquad (3.30\mathrm{b}) \end{aligned}$$

For the data transmission phase, the received signal in (3.6) can be rewritten as

$$\begin{aligned} \mathbf{y}_i^{\mathrm{d}} &= \sum_{k \in \mathcal{U}_i} d_{i,k}^{-\frac{\zeta}{2}} \gamma_{i,k} \mathbf{x}_k^{\mathrm{d}} + \sum_{k \in \mathcal{U}_i^c} d_{i,k}^{-\frac{\zeta}{2}} \gamma_{i,k} \mathbf{x}_k^{\mathrm{d}} + \mathbf{n}_i^{\mathrm{d}} \\ &= \sum_{k \in \mathcal{U}_i} d_{i,k}^{-\frac{\zeta}{2}} \hat{\gamma}_{i,k} \mathbf{x}_k^{\mathrm{d}} + \mathbf{v}_i \\ &= \sum_{k \in \mathcal{U}} d_{i,k}^{-\frac{\zeta}{2}} \hat{\gamma}_{i,k} \mathbf{x}_k^{\mathrm{d}} + \mathbf{v}_i, \ i \in \mathcal{B} \end{aligned} \tag{3.31}$$

where the last step follows from (3.29), and $\mathbf{v}_i$ represents the equivalent interference-plus-noise given as

$$\mathbf{v}_i = \sum_{k\in\mathcal{U}} d_{i,k}^{-\frac{\zeta}{2}}(\gamma_{i,k} - \hat{\gamma}_{i,k})\mathbf{x}_k^{\mathrm{d}} + \mathbf{n}_i^{\mathrm{d}}. \tag{3.32}$$

The correlation of $\{\mathbf{x}_k^{\mathrm{d}}\}$ is given by

$$R_{x_k^{\mathrm{d}}} \triangleq \frac{1}{(1-\alpha)T}\mathbb{E}[\mathbf{x}_k^{\mathrm{d}}(\mathbf{x}_k^{\mathrm{d}})^{\mathrm{H}}] = \beta_k' P_0,\ k \in \mathcal{U}, \tag{3.33a}$$

$$\text{and} \frac{1}{(1-\alpha)T}\mathbb{E}[\mathbf{x}_k^{\mathrm{d}}(\mathbf{x}_m^{\mathrm{d}})^{\mathrm{H}}] = 0,\ \forall k \neq m,\ k, m \in \mathcal{U}. \tag{3.33b}$$

By definition, we obtain

$$\sigma_{v_i}^2 = \frac{1}{(1-\alpha)T}\mathbb{E}[\mathbf{v}_i\mathbf{v}_i^{\mathrm{H}}] = \sum_{k=1}^{K} d_{i,k}^{-\zeta}\beta_k' P_0 \cdot \mathrm{MSE}_{i,k} + N_0,\ i \in \mathcal{B}, \tag{3.34a}$$

$$\text{and} \quad \frac{1}{(1-\alpha)T}\mathbb{E}[\mathbf{v}_i\mathbf{v}_j^{\mathrm{H}}] = 0,\ \forall i \neq j,\ i, j \in \mathcal{B}. \tag{3.34b}$$

We next express (3.31) in a matrix form. Define $\hat{\mathbf{H}}$ as the estimated channel matrix with (i, k)th element given by $\hat{H}_{i,k} = d_{i,k}^{-\frac{\zeta}{2}}\gamma_{i,k}, i \in \mathcal{B}, k \in \mathcal{U}$. Denote by $\mathbf{V} = [\mathbf{v}_1^{\mathrm{T}}, \mathbf{v}_2^{\mathrm{T}}, \cdots, \mathbf{v}_N^{\mathrm{T}}]^{\mathrm{T}}$, $\mathbf{Y}^{\mathrm{d}} = [(\mathbf{y}_1^{\mathrm{d}})^{\mathrm{T}}, (\mathbf{y}_2^{\mathrm{d}})^{\mathrm{T}}, \cdots, (\mathbf{y}_N^{\mathrm{d}})^{\mathrm{T}}]^{\mathrm{T}}$, and $\mathbf{X}^{\mathrm{d}} = [(\mathbf{x}_1^{\mathrm{d}})^{\mathrm{T}}, (\mathbf{x}_2^{\mathrm{d}})^{\mathrm{T}}, \cdots, (\mathbf{x}_N^{\mathrm{d}})^{\mathrm{T}}]^{\mathrm{T}}$. Then

$$\mathbf{Y}^{\mathrm{d}} = \hat{\mathbf{H}}\mathbf{X}^{\mathrm{d}} + \mathbf{V}. \tag{3.35}$$

Note that the interference-plus-noise term $\mathbf{V}$ is in general correlated with the signal part $\hat{\mathbf{H}}\mathbf{X}^{\mathrm{d}}$. Therefore, the achievable rate for (3.35) is lower bounded by the case with independent Gaussian noise [34]. Specifically, the throughput lower bound is given by

$$I(\mathbf{X}^{\mathrm{d}}; \mathbf{Y}^{\mathrm{d}}|\hat{\mathbf{H}}) = \log\det\left(\mathbf{I} + \mathbf{R}_V^{-1}\hat{\mathbf{H}}\mathbf{R}_{X^{\mathrm{d}}}\hat{\mathbf{H}}^{\mathrm{H}}\right) \tag{3.36}$$

where $\mathbf{R}_V = \mathrm{diag}\{\sigma_{v_1}^2, \cdots, \sigma_{v_N}^2\}$ is a diagonal matrix formed by $\{\sigma_{v_n}^2\}$, $\mathbf{R}_{X^{\mathrm{d}}} = \mathrm{diag}\{R_{x_1^{\mathrm{d}}}, \cdots, R_{x_K^{\mathrm{d}}}\}$, and $I(\mathbf{X}^{\mathrm{d}}; \mathbf{Y}^{\mathrm{d}}|\hat{\mathbf{H}})$ is the conditional mutual information between $\mathbf{X}^{\mathrm{d}}$ and $\mathbf{Y}^{\mathrm{d}}$ provided that $\mathbf{x}_k^{\mathrm{d}}$, the k-th row of $\mathbf{X}^{\mathrm{d}}$, is independently drawn from $\mathcal{CN}(0, \beta_k' P_0\mathbf{I})$ for $k = 1, \cdots, K$, and $\mathbf{v}_i$ is independently drawn from $\mathcal{CN}(0, \sigma_{v_i}^2\mathbf{I})$ for $i = 1, \cdots, N$. Considering the two-phase transmission scheme, we obtain the information throughput of the system:

$$R = (1-\alpha)\mathbb{E}\left[\log\det\left(\mathbf{I} + \mathbf{R}_V^{-1}\hat{\mathbf{H}}\mathbf{R}_{X^{\mathrm{d}}}\hat{\mathbf{H}}^{\mathrm{H}}\right)\right]. \tag{3.37}$$

References

1. J. Zhang, X. Yuan and Y. J. Zhang, "Locally Orthogonal Training Design for Cloud-RANs Based on Graph Coloring," *IEEE Transactions on Wireless Communications*, vol. 16, no. 10, pp. 6426–6437, Oct. 2017.
2. A. Checko, H. L. Christiansen, Y. Yan, L. Scolari, G. Kardaras, M. S. Berger, and L. Dittmann, "Cloud RAN for mobile networks - a technology overview," *IEEE Communications Surveys & Tutorials*, vol. 17, no. 1, pp. 405–426, 2015.
3. M. Coldrey (Tapio) and P. Bohlin, "Training-based MIMO systems-part I: performance comparison," *IEEE Transactions on Signal Processing*, vol. 55, no. 11, pp. 5464–5476, Nov. 2007.
4. X. Yuan, C. Fan, and Y. J. Zhang, "Fundamental limits of training-based multiuser MIMO systems," available at: http://arxiv.org/abs/1511.08977.
5. E. Malaguti, M. Monaci, and P. Toth, "A metaheuristic approach for the vertex coloring problem," *INFORMS Journal on Computing*, vol. 20, no. 2, pp. 302–316, 2008.
6. D. Brelaz, "New methods to color the vertices of a graph," *Communications of the ACM*, vol. 22, pp. 251–256, 1979.
7. T. L. Marzetta, "Noncooperative cellular wireless with unlimited numbers of base station antennas," *IEEE Transactions on Wireless Communications*, vol. 9, no. 11, pp. 3590–3600, Nov. 2010.
8. F. Fernandes, A. Ashikhmin, and T. L. Marzetta, "Inter-cell interference in noncooperative TDD large scale antenna systems," *IEEE Journal on Selected Areas in Communications*, vol. 31, no. 2, pp. 192–201, Feb. 2013.
9. X. Xie, M. Peng, W. Wang, and H. V. Poor, "Training design and channel estimation in uplink cloud radio access networks," *IEEE Signal Processing Letters*, vol. 22, no. 8, pp. 1060–1064, Aug. 2015.
10. J. Ma, and L. Ping, "Data-aided channel estimation in large antenna systems," *IEEE Transactions on Signal Processing*, vol. 62, no. 12, pp. 3111–3124, Jun. 2014.
11. K. Upadhya, S. A. Vorobyov, and M. Vehkapera, "Superimposed pilots are superior for mitigating pilot contamination in massive MIMO" Part I: theory and channel estimation," Online, available at: http://arxiv.org/pdf/1603.00648v1.pdf.
12. H. Q. Ngo, and E. Larsson, "Evd-based channel estimation in multi-cell multiuser MIMO systems with very large antenna arrays," in *Proc. of IEEE International Conference on Acoustics, Speech and Signal Processing (ICASSP)*, 2012, pp. 3249–3252.
13. R. R. Muller, L. Cottatellucci, and M. Vehkapera, "Blind pilot decontamination," *IEEE Journal on Selected Areas in Communications*, vol. 8, no. 5, pp. 773–786, Mar. 2014.
14. H. Huh, G. Caire, H. Papadopoulos, and S. Ramprashad, "Achieving massive MIMO spectral efficiency with a not-so-large number of antennas," *IEEE Transactions on Wireless Communications*, vol. 11, no. 9, pp.3226–3239, Sep. 2012.
15. X. Zhu, Z. Wang, C. Qian, L. Dai, J. Chen, S. Chen, and L. Hanzo, "Soft pilot reuse and multi-cell block diagonalization precoding for massive MIMO systems," *IEEE Transactions on Vehicular Technology*, vol. PP, no. 99, pp. 1–1, 2015.
16. H. Yang, and T. L. Marzetta, "Performance of pilot reuse in multi-cell massive MIMO," in *Proc. of IEEE BlackSeaCom'15*, Constanta, 2015.
17. H. Yin, D. Gesbert, M. Filippou, and Y. Liu, "A coordinated approach to channel estimation in large-scale multiple-antenna systems," *IEEE Journal on Selected Areas in Communications*, vol. 31, no. 2, pp. 264–273, Feb. 2013.
18. O. Y. Bursalioglu, C. Wang, H. Papadopoulos, and G. Caire, "RRH based massive MIMO with "on the Fly" pilot contamination control," Online, available at: http://arxiv.org/pdf/1601.01983v1.pdf.
19. C. Li, J. Zhang, S. Song, and K. B. Letaief, "Selective uplink training for massive MIMO systems," Online, available at: http://arxiv.org/pdf/1602.08857v1.pdf.

20. H. Ahmadi, A. Farhang, N. Marchetti, and A. MacKenzie, "A game theoretical approach for pilot contamination avoidance in massive MIMO," *IEEE Wireless Communications Letters*, vol. 5, no. 1, pp. 12–15, Feb. 2016.
21. J. Jose, A. Ashikhmin, T. Marzetta, and S. Vishwanath, "Pilot contamination and precoding in multi-cell TDD systems," *IEEE Transactions on Wireless Communications*, vol. 10, no. 8, pp. 2640–2651, Aug. 2011.
22. 3GPP TR 36.817 v10.0.0, "Evolved Universal Terrestrial Radio Access (E-UTRA); Uplink multiple antenna transmission; Base Station (BS) radio transmission and reception (Release 10)."
23. 3GPP TR 36.871 v11.0.0, "Evolved Universal Terrestrial Radio Access (E-UTRA); Downlink Multiple Input Multiple Output (MIMO) enhancement for LTE-Advanced (Release 11)."
24. A. Liu and V. Lau, "Joint power and antenna selection optimization in large cloud radio access networks," *IEEE Transactions on Signal Processing*, vol. 62, no. 5, pp. 1319–1328, Mar. 2014.
25. Y. Shi, J. Zhang, and K. Letaief, "Group sparse beamforming for green cloud-RAN," *IEEE Transactions on Wireless Communications*, vol. 13, no. 5, pp. 2809–2823, May 2014.
26. C. Fan, Y. Zhang, and X. Yuan, "Dynamic nested clustering for parallel PHY-layer processing in cloud-RAN," *IEEE Transactions on Wireless Communications*, vol. PP, no. 99, pp. 1–1, 2015.
27. M. R. Garey, and D. S. Johnson, *Computers and intractability: a guide to the theory of NP-completeness*, W. H. Freeman & Co, New York. 1979.
28. B. N. Clark, C. J. Colbourn, and D. S. Johnson, "Unit disk graphs," *Discrete Mathematics*, vol. 86, pp. 165–177, 1990.
29. I.-M. Diaz, P. Zabala, "A branch-and-cut algorithm for graph coloring," *Discrete Applied Mathematics*, vol. 154, no. 5, pp. 826–847, 2006.
30. T. J. Sager, S. Lin, "A pruning procedure for exact graph coloring," *ORSA Journal on Computing*, vol. 3, pp.226–230, 1991.
31. M. J. B. Appel and R. P. Russo, "The maximum vertex degree of a graph on uniform points in $[0, 1]^d$", *Advances in Applied Probability,* vol. 29, pp. 567–581, 1997.
32. C. J. H. McDiarmid and T. Muller, "Colouring random geometric graphs," in *Proc. of European Conference on Combinatorics, Graph Theory and Applications*, pp. 1–4, 2005.
33. S. M. Kay, *Fundamentals of Statistical Signal Processing*, Prentice Hall, 2001.
34. M. Medard, "The effect upon channel capacity in wireless communications of perfect and imperfect knowledge of the channel," *IEEE Transactions on Information Theory*, vol. 46, no. 3, pp. 933–946, May 2000.
35. M. D. Penrose, *Random Geometric Graphs*, Oxford University Press, Oxford 2003.

Chapter 4
Scalable Signal Detection: Dynamic Nested Clustering

In Chap. 2, we proposed a threshold-based channel sparsification approach, and showed that the channel matrices can be greatly sparsified without substantially compromising the system capacity. In this chapter and the next chapter, we endeavor to design scalable algorithms for joint signal detection in the uplink of C-RAN by exploiting the high sparsity of the channel matrix. In this chapter, we propose a dynamic nested clustering (DNC) algorithm which greatly reduces the computational complexity of MMSE detection from $O(N^3)$ to $O(N^a)$, where N is the total number of RRHs and $a \in (1, 2]$ is a constant determined by the computation implementations. In the next chapter, we propose a randomized Gaussian message passing (RGMP) algorithm, which further reduces the computational complexity of MMSE detection to be linear in the number of RRHs. The material in this chapter is mainly based on [1].

4.1 System Model and Problem Formulation

In this chapter, we consider the uplink transmission of a C-RAN with N single-antenna RRHs, and K single-antenna mobile users uniformly located over the entire coverage area. The received signal vector $\mathbf{y} \in \mathbb{C}^{N\times 1}$ at the RRHs is

$$\mathbf{y} = \mathbf{H}\mathbf{P}^{\frac{1}{2}}\mathbf{x} + \mathbf{n}, \tag{4.1}$$

where $\mathbf{H} \in \mathbb{C}^{N\times K}$ denotes the channel matrix, with the (n, k)th entry $H_{n,k}$, being the channel coefficient between the kth user and the nth RRH. $\mathbf{P} \in \mathbb{R}^{N\times N}$ is a diagonal matrix with the kth diagonal entry P_k being the transmitting power allocated to user k. $\mathbf{x} \in \mathbb{C}^{K\times 1}$ is the vector of the transmitted signal from the K users and $\mathbf{n} \sim \mathcal{CN}(\mathbf{0}, N_0\mathbf{I})$ is the vector of noise received by RRHs. The transmit

Y.-J. A. Zhang et al., *Scalable Signal Processing in Cloud Radio Access Networks*, SpringerBriefs in Electrical and Computer Engineering,
https://doi.org/10.1007/978-3-030-15884-2_4

signals are assumed to follow an independent complex Gaussian distribution with unit variance, i.e. $E[\mathbf{x}\mathbf{x}^{\mathrm{H}}] = \mathbf{I}$. Specifically, $H_{n,k} = \gamma_{n,k} d_{n,k}^{-\frac{\zeta}{2}}$, where $\gamma_{n,k}$ is the i.i.d. Rayleigh fading coefficient with zero mean and variance 1, $d_{n,k}$ is the distance between the nth RRH and the kth user, and ζ is the path loss exponent. Then, $d_{n,k}^{-\zeta}$ is the path loss from the kth user to the nth RRH. In this chapter, to reduce the computational complexity of signal detection, we employ the threshold-based channel sparsification approach proposed in Chap. 2 to sparsify the channel matrix $\mathbf{H}$. That is,

$$\widehat{H}_{n,k} = \begin{cases} H_{n,k}, & d_{n,k} < d_0 \\ 0, & \text{otherwise.} \end{cases} \tag{4.2}$$

Based on the sparsified channel matrix $\widehat{\mathbf{H}}$, we employ linear MMSE detection to estimate the transmitted signal vector $\mathbf{x}$, with the decision statistics given by

$$\widehat{\mathbf{x}} \approx \mathbf{P}^{\frac{1}{2}} \widehat{\mathbf{H}}^{\mathrm{H}} \left(\widehat{\mathbf{H}} \mathbf{P} \widehat{\mathbf{H}}^{\mathrm{H}} + \widehat{N}_0 \mathbf{I} \right)^{-1} \mathbf{y}, \tag{4.3}$$

where we assume that the transmission power $\widehat{N}_0 = \mathrm{E}[\sum_{j \neq k} P_j |\widetilde{H}_{n,j}|^2] + N_0$ for arbitrary RRH n. In the above, the inversion of the $N \times N$ matrix $\widehat{\mathbf{A}}$ requires computational complexity of $O(N^3)$. This complexity is prohibitively high for a large-scale C-RAN with hundreds and thousands of RRHs, thus posing a serious scalability problem. In what follows, we endeavour to design a low-complexity algorithm to estimate $\mathbf{x}$ by MMSE detection, where the complexity is kept at a tolerable level even for a network with very large size. In particular, our work is divided into the following two steps.

- We show that by skillfully indexing the RRHs, the sparsified channel matrix has a (nested) doubly bordered block diagonal (DBBD) structure, as shown in Fig. 4.1. Interestingly, we find that the DBBD matrix naturally leads to a dynamic nested clustering (DNC) algorithm that greatly improves the scalability in terms of the system. Specifically, the diagonal blocks (see Fig. 4.1) can be interpreted as clusters (or sub-networks) that are processed independently. Different clusters are coordinated by the cut-node block and border blocks that capture the interference among clusters. As such, the baseband-processing complexity is dominated by the size of the clusters instead of the entire C-RAN network.
- Thanks to the centralized BBU pool of C-RAN, the DNC algorithm is amenable for different processing implementations through adjusting the size and the number of the clusters. We design different clustering strategies for both serial processing and parallel processing to minimize the computational complexity and the computation time, respectively.

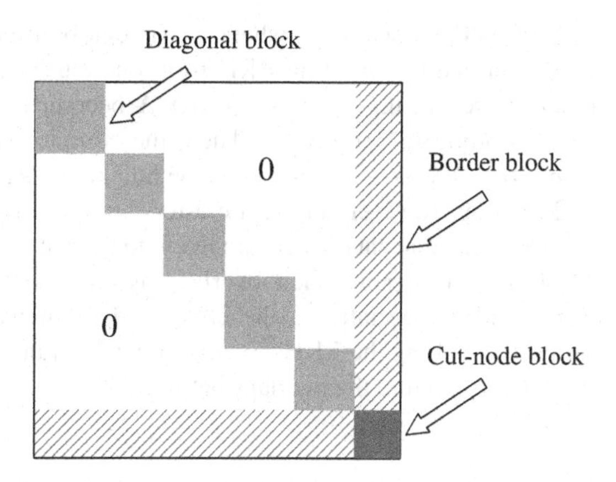

Fig. 4.1 A matrix in a doubly bordered block diagonal form

4.2 Single-Layer Dynamic Nested Clustering

In this section, we present the single-layer DNC algorithm based on the sparsified channel matrix. As shown in (4.3), estimating $\mathbf{x}$ is equivalent to calculating $\mathbf{P}^{\frac{1}{2}}\widehat{\mathbf{H}}^{\mathrm{H}}\widehat{\mathbf{A}}^{-1}\mathbf{y}$, where

$$\widehat{\mathbf{A}} = \widehat{\mathbf{H}}\mathbf{P}\widehat{\mathbf{H}}^{\mathrm{H}} + \widehat{N}_0\mathbf{I}. \tag{4.4}$$

The computational complexity is in general $O(N^3)$, and is dominated by the calculation of $\widehat{\mathbf{A}}^{-1}\mathbf{y}$. Nonetheless, the average number of non-zero entries in each row of $\widehat{\mathbf{H}}$ is a constant $\xi = \pi d_0^2 \beta_K$. The computational complexity of multiplying $\mathbf{P}^{\frac{1}{2}}\widehat{\mathbf{H}}^{\mathrm{H}}$ and $\widehat{\mathbf{A}}^{-1}\mathbf{y} \in \mathbb{C}^{N\times 1}$ is only $O(\xi N)$ and is much smaller than $O(N^3)$. Therefore, we focus on reducing the computational complexity of calculating $\widehat{\mathbf{A}}^{-1}\mathbf{y}$, which is equivalent to solving the following system of $\boldsymbol{\Omega}$:

$$\widehat{\mathbf{A}}\boldsymbol{\Omega} = \mathbf{y}. \tag{4.5}$$

Clearly, $\widehat{\mathbf{A}}$ is a sparse matrix. Solving sparse linear equations have been studied in various areas, such as numerical linear algebra, graph theory, etc. However, existing work mostly focuses on iterative algorithms, with no guarantee on accuracy and convergence [2]. Here, we propose a non-iterative algorithm that yields an accurate solution of (4.5). Before delving into details, we first explain the physical meanings of the entries in $\widehat{\mathbf{A}}$ in the following. According to the threshold-based channel sparsification, the (n, k)th entry of channel matrix $\widehat{\mathbf{H}}$ is non-zero only when the kth user is in the service region of RRH n (defined as a circular area of radius d_0 centered around RRH n). Consequently, from the definition of $\widehat{\mathbf{A}}$ in (4.4), the (n_1, n_2)th entry in $\widehat{\mathbf{A}}$ is non-zero only when the service regions of RRH n_1 and n_2 overlap each other, and there is at least one user falling into the overlapped region.

Consider an ideal case where RRHs can be divided into disjoint clusters, that is, the service region of an RRH from one cluster does not overlap that from any other cluster. In this case, the matrix $\widehat{\mathbf{A}}$ becomes block-diagonal with each block corresponding to one cluster. Then, the complexity of calculating $\widehat{\mathbf{A}}^{-1}\mathbf{y}$ reduces from $O(N^3)$ to $O(n^3)$, where n is the number of RRHs in a cluster.

In reality, however, adjacent clusters are not disjoint, i.e., the service regions of the RRHs in adjacent clusters are likely to overlap. Traditional clustering algorithms [3, 4] usually ignore such overlapping, resulting in a noticeable performance degradation, especially in the cluster edge. In what follows, we show that by properly labelling the RRHs, matrix $\widehat{\mathbf{A}}$ can be transformed to a DBBD form, where the borders capture the overlaps between clusters.

4.2.1 RRH Labelling Algorithm

To start with, we give the definition of a Hermitian DBBD matrix as follows:

Definition 4.1 A matrix $\mathbf{A}$ is said to be a Hermitian DBBD matrix if it is in the following form

$$\mathbf{A} = \begin{bmatrix} \mathbf{A}_{1,1} & \mathbf{0} & \cdots & \mathbf{0} & \mathbf{A}_{c1}^{\mathrm{H}} \\ \mathbf{0} & \mathbf{A}_{2,2} & \ddots & \vdots & \mathbf{A}_{c2}^{\mathrm{H}} \\ \vdots & \ddots & \ddots & \mathbf{0} & \vdots \\ \mathbf{0} & \cdots & \mathbf{0} & \mathbf{A}_{m,m} & \mathbf{A}_{cm}^{\mathrm{H}} \\ \mathbf{A}_{c1} & \mathbf{A}_{c2} & \cdots & \mathbf{A}_{cm} & \mathbf{A}_{c} \end{bmatrix}, \tag{4.6}$$

where the diagonal blocks $\mathbf{A}_{ii}$ are $n_i \times n_i$ Hermitian matrices, the border blocks $\mathbf{A}_{ci}$ are $n_c \times n_i$ matrices, and the cut-node block $\mathbf{A}_c$ is an $n_c \times n_c$ Hermitian matrix.

We divide the entire C-RAN area into disjoint sub-areas as illustrated in Fig. 4.2. We then further divide each sub-area into a width-d_0 boundary and a center area which are colored by white and grey, respectively. Recall that d_0 is the distance threshold used in the channel sparsification. Note that the RRHs in a center area do not have overlapped service regions with those in other center areas. Only RRHs in the width-d_0 boundary may have overlapped service regions with the RRHs in adjacent sub-areas. This implies that matrix $\widehat{\mathbf{A}}$ can be transformed to a DBBD matrix with each diagonal block corresponding to the RRHs in a center area and the cut-node block corresponding to the RRHs in the width-d_0 boundaries. The border blocks of $\widehat{\mathbf{A}}$ capture the interaction between different clusters through the cut-node block.

Notice that the sub-areas can be in any shape, such as rectangle, square, hexagon and so on. In this book, without loss of generality, we use squares with side length

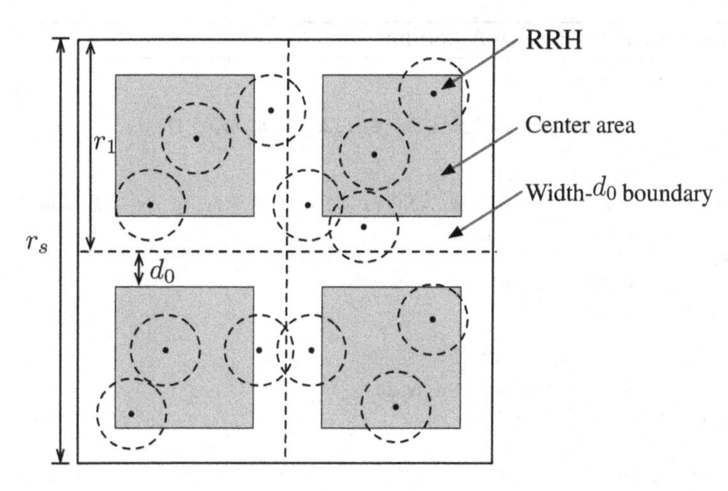

Fig. 4.2 Geographical RRH grouping in C-RAN

r_1 as an example. The RRH labelling algorithm is given in Algorithm 3.1, where $b(n)$ is the label of RRH n. The coordinates of RRH n are denoted as:

$$l_n = (lx_n, ly_n), \tag{4.7}$$

where $lx_n \in [0, a_x], ly_n \in [0, a_y]$. a_x and a_y are the side lengths of the whole network. In steps 2–9 of Algorithm 3.1, we first divide the overall network into disjoint squares with side length r_1, and then group the RRHs into center clusters or the boundary cluster according to their locations. In steps 10–18, the RRHs are numbered based on the cluster they belong to. After numbering all the RRHs, we organize the matrix $\widehat{\mathbf{A}}$ and the signal vector $\mathbf{y}$ in the ascending order of the RRHs' numbers. For example, the first row of $\widehat{\mathbf{A}}$ corresponds to the RRH with label $b(n) = 1$. The matrix $\widehat{\mathbf{A}}$ now becomes a DBBD matrix.

Remark 4.1 We would like to emphasize that RRH labelling in Algorithm 3.1 is based solely on the locations of RRHs instead of the instantaneous CSI. Thus, RRH labelling only needs to be conducted once in system initialization. That means the complexity of RRH labelling is negligible compared with that of the MMSE detection.

4.2.2 Single-Layer DNC

Now that $\widehat{\mathbf{A}}$ is converted into a DBBD matrix, we are ready to present the DNC algorithm. Suppose that the DBBD matrix $\widehat{\mathbf{A}}$ has m_1 diagonal blocks. Then, (4.5) becomes

Algorithm 4.1 RRH Labelling Algorithm

Require: $a_x, a_y, d_0, r_1, l_n, \forall n$
Ensure: $b(n), \forall n$
1: Set $m_x = \lceil \frac{a_x}{r_1} \rceil$, $m_y = \lceil \frac{a_y}{r_1} \rceil$ and $\mathcal{C}_i = \varnothing, \forall i \in \{1, 2, \cdots, m_x m_y + 1\}$
2: **for** $n = 1$ to N **do**
3: Setting $i = \lceil \frac{lx_n}{r_1} \rceil$, $j = \lceil \frac{ly_n}{r_1} \rceil$
4: **if** $(i-1)r_1 + d_0 \leq lx_n \leq ir_1 - d_0$ AND $(j-1)r_1 + d_0 \leq ly_n \leq jr_1 - d_0$ **then**
5: $\mathcal{C}_{(i-1)m_x+j} \leftarrow \{\mathcal{C}_{(i-1)m_x+j}, n\}$
6: **else**
7: $\mathcal{C}_{m_x m_y+1} \leftarrow \{\mathcal{C}_{m_x m_y+1}, n\}$
8: **end if**
9: **end for**
10: Set $j = 1$
11: **for** $i = 1$ to $m_x m_y + 1$ AND $n = 1$ to N **do**
12: **if** $n \in \mathcal{C}_i$ **then**
13: Label RRH n by j: $b(n) \leftarrow j$
14: $j \leftarrow j + 1$
15: **end if**
16: **end for**

$$\begin{bmatrix} \widehat{\mathbf{A}}_{1,1} & \mathbf{0} & \cdots & \mathbf{0} & \widehat{\mathbf{A}}_{c1}^{\mathrm{H}} \\ \mathbf{0} & \widehat{\mathbf{A}}_{2,2} & \ddots & \vdots & \widehat{\mathbf{A}}_{c2}^{\mathrm{H}} \\ \vdots & \ddots & \ddots & \mathbf{0} & \vdots \\ \mathbf{0} & \cdots & \mathbf{0} & \widehat{\mathbf{A}}_{m_1,m_1} & \widehat{\mathbf{A}}_{cm_1}^{\mathrm{H}} \\ \widehat{\mathbf{A}}_{c1} & \widehat{\mathbf{A}}_{c2} & \cdots & \widehat{\mathbf{A}}_{cm_1} & \widehat{\mathbf{A}}_{c} \end{bmatrix} \begin{bmatrix} \boldsymbol{\Omega}_1 \\ \boldsymbol{\Omega}_2 \\ \vdots \\ \boldsymbol{\Omega}_{m_1} \\ \boldsymbol{\Omega}_c \end{bmatrix} = \begin{bmatrix} \mathbf{y}_1 \\ \mathbf{y}_2 \\ \vdots \\ \mathbf{y}_{m_1} \\ \mathbf{y}_c \end{bmatrix}, \tag{4.8}$$

where the $n_i \times 1$ vectors $\boldsymbol{\Omega}_i$ and $\mathbf{y}_i$ are sub-vectors of $\boldsymbol{\Omega}$ and $\mathbf{y}$, respectively. Likewise, $\boldsymbol{\Omega}_c$ and $\mathbf{y}_c$ are $n_c \times 1$ sub-vectors.

The solution to the above equation is given by

$$\boldsymbol{\Omega}_c = \left(\widehat{\mathbf{A}}_c - \sum_{i=1}^{i=m_1} \widehat{\mathbf{A}}_{ci} \widehat{\mathbf{A}}_{i,i}^{-1} \widehat{\mathbf{A}}_{ci}^{\mathrm{H}} \right)^{-1} \left(\mathbf{y}_c - \sum_{i=1}^{i=m_1} \widehat{\mathbf{A}}_{ci} \widehat{\mathbf{A}}_{i,i}^{-1} \mathbf{y}_i \right), \tag{4.9}$$

and

$$\boldsymbol{\Omega}_i = \widehat{\mathbf{A}}_{i,i}^{-1} \left(\mathbf{y}_i - \widehat{\mathbf{A}}_{ci}^{\mathrm{H}} \boldsymbol{\Omega}_c \right), \tag{4.10}$$

for all $i \in \{1, 2, \cdots, m_1\}$. From (4.9), we note that $\boldsymbol{\Omega}_c$, the sub-vector corresponding to the cut-node block, can be calculated independently of the other sub-vectors $\boldsymbol{\Omega}_i$. Moreover, after $\boldsymbol{\Omega}_c$ is obtained from (4.9), the calculation of $\boldsymbol{\Omega}_i$ as given by (4.10) only involves the ith diagonal block of $\widehat{\mathbf{A}}$ and the corresponding ith border block $\widehat{\mathbf{A}}_{ci}$. That is, the complexity of calculating each $\boldsymbol{\Omega}_i$ is dominated by the sizes of $\widehat{\mathbf{A}}_{i,i}$ and $\widehat{\mathbf{A}}_{ci}$ instead of the entire matrix $\widehat{\mathbf{A}}$. This implies that the complexity of calculating $\boldsymbol{\Omega}$ is reduced. In addition, as the calculation of $\boldsymbol{\Omega}_i$ is independent from each other, we can calculate each $\boldsymbol{\Omega}_i$ in parallel to further reduce the total computation time (i.e., the length of time of calculating $\boldsymbol{\Omega}$). In the following subsection, we will derive the lowest computational complexity of the proposed single-layer DNC algorithm. Then, in Sect. 4.2.4, we will give a parallel implementation of the single-layer DNC algorithm.

4.2.3 Optimizing the Computational Complexity

Table 4.1 lists the detailed computational complexity of each step in the single-layer DNC algorithm. In Table 4.1, $N_{d,1}$ and $N_{b,1}$ are the average sizes of the diagonal blocks and cut-node block respectively. m_1 is the average number of diagonal blocks. ξ_1 is the average number of non-zero entries per row of $\widehat{\mathbf{A}}$. In practice, ξ_1 is roughly a constant independent of the network size N. The total computational complexity of the single-layer DNC algorithm is $O(N_{b,1}^3 + (N_{d,1}^3 + N_{b,1}^2)m_1)$.

To achieve a lower computational complexity, $N_{d,1}$, $N_{b,1}$ and m_1 should be as small as possible. However, the block sizes and the number of blocks cannot be adjusted arbitrarily. In fact, for a given r_1, there is a fixed ratio between $N_{d,1}$ and $N_{b,1}$. We denote this ratio as $N^{z_1} = \frac{N_{d,1}}{N_{b,1}}$. Specifically, when the C-RAN is a square with side length r_s, the relationship between r_1 and the ratio N^{z_1} is

$$(r_1 - 2d_0)^2 = 4(r_1 - d_0)d_0\frac{r_s^2}{r_1^2}N^{z_1}. \tag{4.11}$$

Table 4.1 Computational complexity of each step in evaluating (4.9) and (4.10)

Step	Operation	Complexity/operation	Total number of operations
1	$\widehat{\mathbf{A}}_{i,i}^{-1}$	$O(N_{d,1}^3)$	m_1
2	$\widehat{\mathbf{A}}_{ci}\widehat{\mathbf{A}}_{i,i}^{-1}\widehat{\mathbf{A}}_{ci}^{\mathrm{H}}$ $\widehat{\mathbf{A}}_{ci}\widehat{\mathbf{A}}_{i,i}^{-1}\mathbf{y}_i$	$O(\frac{\xi_1}{m_1}N_{b,1}N_{d,1})$	m_1
3	$\widehat{\mathbf{A}}_c - \sum_{i=1}^{i=m_1}\widehat{\mathbf{A}}_{ci}\widehat{\mathbf{A}}_{i,i}^{-1}\widehat{\mathbf{A}}_{ci}^{\mathrm{H}}$ $\mathbf{y}_c - \sum_{i=1}^{i=m_1}\widehat{\mathbf{A}}_{ci}\widehat{\mathbf{A}}_{i,i}^{-1}\mathbf{y}_i$	$O(m_1N_{b,1}^2)$	1
4	$\left(\widehat{\mathbf{A}}_c - \sum_{i=1}^{i=m_1}\widehat{\mathbf{A}}_{ci}\widehat{\mathbf{A}}_{i,i}^{-1}\widehat{\mathbf{A}}_{ci}^{\mathrm{H}}\right)^{-1}$ $\times\left(\mathbf{y}_c - \sum_{i=1}^{i=m_1}\widehat{\mathbf{A}}_{ci}\widehat{\mathbf{A}}_{i,i}^{-1}\mathbf{y}_i\right)$	$O(N_{b,1}^3)$	1
5	$\mathbf{y}_i - \widehat{\mathbf{A}}_{ci}^{\mathrm{H}}\boldsymbol{\Omega}_c$	$O(\frac{\xi_1}{m_1}N_{b,1})$	m_1
6	$\widehat{\mathbf{A}}_{i,i}^{-1}(\mathbf{y}_i - \widehat{\mathbf{A}}_{ci}^{\mathrm{H}}\boldsymbol{\Omega}_c)$	$O(N_{d,1}^2)$	m_1

Without loss of generality, we assume that both $N_{d,1}$ and $N_{b,1}$ are at least 1. Otherwise, clustering will become meaningless as we still need to do matrix inversion over the entire matrix $\widehat{\mathbf{A}}$. Then, by adjusting r_1 from $2d_0$ to r_s, z_1 goes from -1 to 1. Before deriving $N_{d,1}$, $N_{b,1}$ and m_1, we first introduce the definition of Little-o, which is often written as $o(\cdot)$. For an arbitrary function $f(x)$, if $\lim_{x\to\infty}\frac{f(x)}{\phi(x)}=0$, then it is said that $f(x)\in o(\phi(x))$, or $f(x)=o(\phi(x))$. Then, based on (4.11), we obtain the following results on $N_{d,1}$, $N_{b,1}$ and m_1:

Lemma 4.1 *In large square C-RANs, $N_{d,1}$, $N_{b,1}$ and m_1 are*

$$N_{d,1}=\left(4d_0\beta_N^{\frac{1}{2}}N^{1+z_1}\right)^{\frac{2}{3}}+o(N^{\frac{2}{3}z_1+\frac{2}{3}}), \tag{4.12}$$

$$N_{b,1}=\left(4d_0\beta_N^{\frac{1}{2}}N^{1-\frac{z_1}{2}}\right)^{\frac{2}{3}}+o(N^{-\frac{1}{3}z_1+\frac{2}{3}}), \tag{4.13}$$

$$\lim_{r_s\to\infty} m_1=(4d_0)^{-\frac{2}{3}}\beta_N^{-\frac{1}{3}}N^{\frac{1}{3}-\frac{2}{3}z_1} \tag{4.14}$$

where β_N is the RRH density.

Proof r_1 is the solution of (4.11), and we have

$$(4d_0r_s^2N^{z_1})^{\frac{1}{3}}\le r_1\le(4d_0r_s^2N^{z_1})^{\frac{1}{3}}+2d_0, \tag{4.15}$$

Then, since d_0 is a constant as r_s goes to infinity, we can write r_1 as

$$r_1=(4d_0r_s^2N^{z_1})^{\frac{1}{3}}+o(r_s^{\frac{2}{3}}N^{\frac{z_1}{3}}). \tag{4.16}$$

Thus,

$$N_{d,1}=\beta_N(r_1-2d_0)^2=\left(4d_0\beta_N^{\frac{1}{2}}N^{1+z_1}\right)^{\frac{2}{3}}+o(N^{\frac{2}{3}z_1+\frac{2}{3}}), \tag{4.17}$$

$$N_{b,1}=N_{d,1}N^{-z_1}=\left(4d_0\beta_N^{\frac{1}{2}}N^{1-\frac{z_1}{2}}\right)^{\frac{2}{3}}+o(N^{-\frac{1}{3}z_1+\frac{2}{3}}). \tag{4.18}$$

Moreover, as $\lim_{r_s\to\infty}\frac{r_1}{r_s}=(4d_0)^{\frac{1}{3}}\beta_N^{\frac{1}{6}}N^{\frac{1}{3}z_1-\frac{1}{6}}$, we have

$$\lim_{r_s\to\infty} m_1=\lim_{r_s\to\infty}\frac{r_s^2}{r_1^2}=(4d_0)^{-\frac{2}{3}}\beta_N^{-\frac{1}{3}}N^{\frac{1}{3}-\frac{2}{3}z_1}. \tag{4.19}$$

Recall that the computational complexity of the single-layer DNC algorithm is $O(N_{b,1}^3 + (N_{d,1}^3 + N_{b,1}^2)m_1)$. Then, by ignoring $o(N^{\frac{2}{3}z_1+\frac{2}{3}})$, $o(N^{-\frac{1}{3}z_1+\frac{2}{3}})$ and d_0, β_N, ξ_1, we obtain the optimal computational complexity in Proposition 4.1.

Proposition 4.1 *The lowest computational complexity of the single-layer DNC algorithm is* $O(N^{\frac{15}{7}})$ *with the optimal* $z_1 = -\frac{1}{7}$.

4.2.4 Parallel Computing

In this subsection, we first show the detailed parallel implementation of the single-layer DNC algorithm. Then, we try to minimize the total computation time of the parallel single-layer DNC algorithm.

As mentioned in Sect. 4.2.2, we can calculate each $\mathbf{\Omega}_i$ in parallel as the calculation of $\mathbf{\Omega}_i$ is independent from each other. In other words, if we treat each diagonal block as a cluster, then the signals received by each cluster can be processed in parallel of each other. Meanwhile the interactions between different clusters are captured by Ω_c and the border blocks. Figure 4.3 shows how parallel signal processing is deployed in a C-RAN BBU pool. The arrows in Fig. 4.3 indicate the data flows between the processing units. As the figure shows, to expedite the calculation of Ω_c, matrices $\widehat{\mathbf{A}}_{ci}\widehat{\mathbf{A}}_{i,i}^{-1}\widehat{\mathbf{A}}_{ci}^{\mathrm{H}}$ and vectors $\widehat{\mathbf{A}}_{ci}\widehat{\mathbf{A}}_{i,i}^{-1}\mathbf{y}_i$ are calculated at the same time by a number of parallel processing units, and then fed into the central processing unit. Then, Ω_c is calculated in the central processing unit. The result is fed back to the parallel processing units, which will then calculate Ω_i in parallel. That is, steps 1 and 2 in Table 4.1 are first carried out in the parallel processing units. After receiving the results of steps 1 and 2, the central processing unit performs steps 3 and 4. At last, steps 5 and 6 are carried out in the parallel processing units.

Now, let us minimize the total computation time of the single-layer DNC algorithm, i.e., the length of time required to perform the algorithm. Suppose that all the processing units have the same processing power. Without loss of generality,

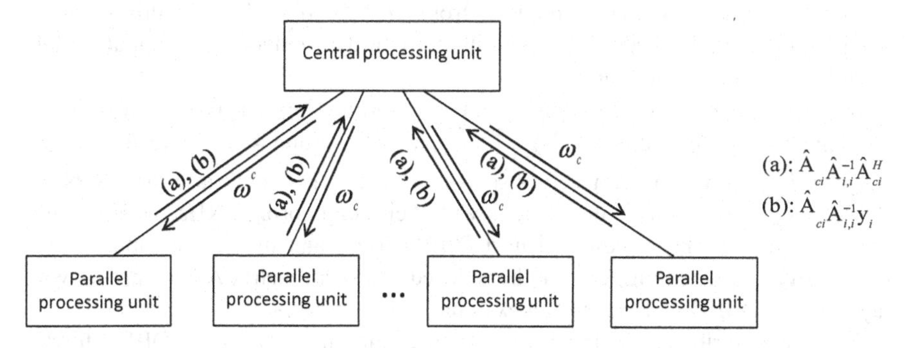

Fig. 4.3 Architecture of parallel computing of the single-layer DNC algorithm

the processing power of a parallel processing unit is normalized to be 1. That is, the computation time to process an operation with complexity $O(N^x)$ is $O(N^3)$. As the operations in steps 1, 2, 5 and 6 can be performed in parallel, the total computation time is $O(N_{d,1}^3+N_{b,1}^3+m_1 N_{b,1}^2)$ according to Table 4.1. Then, based on Lemma 4.1, we obtain the optimal computation time in Proposition 4.2.

Proposition 4.2 *Consider a C-RAN which has processing units with equal processing power. The optimal computation time, $O(N^2)$, is achieved with $z_1 = 0$.*

Remark 4.2 So far, we have assumed that there are always enough parallel processing units, regardless of r_1 or z_1. In this case, we only need to optimize the sizes of diagonal blocks and the cut-node block and ignore the number of blocks. Instead, when the number of parallel processing units is limited, the number of blocks, m_1, also has an effect on the total computation time. Based on Lemma 4.1, m_1 can also be adjusted by r_1 or z_1. In this way, we can balance the computation time with limited availability of processing units. More detailed analysis, however, is out of the scope of this book.

Remark 4.3 In this book, we have only analysed the case where all the processing units have equal processing power. However, we would like to emphasize that the proposed DNC algorithm is adaptive to various data center architectures. For example, when the processing power of the processing units is not equal, it is natural to allocate more computations to the processing units with more processing power. Our proposed DNC algorithm allows flexible computation allocation by easily adjusting both the size and the number of the center clusters as well as the size of the boundary cluster.

4.3 Multi-Layer DNC Algorithm

In the preceding section, we propose a single-layer DNC algorithm, to reduce the computational complexity from $O(N^3)$ to $O(N^{\frac{15}{7}})$. Moreover, when parallel computing is deployed and all the processing units have the same processing power, the total computation time is reduced from $O(N^3)$ to $O(N^2)$. In this section, we propose a multi-layer DNC algorithm to further reduce the computational complexity and computation time.

We notice that the computational complexity of the single-layer DNC algorithm is dominated by calculating $(\widehat{\mathbf{A}}_c - \sum_{i=1}^{i=m_1} \widehat{\mathbf{A}}_{ci}\widehat{\mathbf{A}}_{i,i}\widehat{\mathbf{A}}_{ci}^{\mathrm{H}})^{-1}$ and $\widehat{\mathbf{A}}_{i,i}^{-1}$. Interestingly, the diagonal blocks $\widehat{\mathbf{A}}_{i,i}$ are themselves sparse matrices. This is because the RRHs in the same cluster only have interactions with their neighboring RRHs. This implies that $\widehat{\mathbf{A}}_{i,i}$ can also be represented in a DBBD form, and thus the computational complexity of calculating $\widehat{\mathbf{A}}_{i,i}^{-1}$ can be reduced. As such, matrix $\widehat{\mathbf{A}}$ becomes a two-layer nested DBBD matrix, with an example shown in Fig. 4.4.

We now describe the RRH labelling strategy that turns $\widehat{\mathbf{A}}_{i,i}$ into a DBBD form. We take the top-left cluster in Fig. 4.5 as an example. For this cluster, the RRHs in

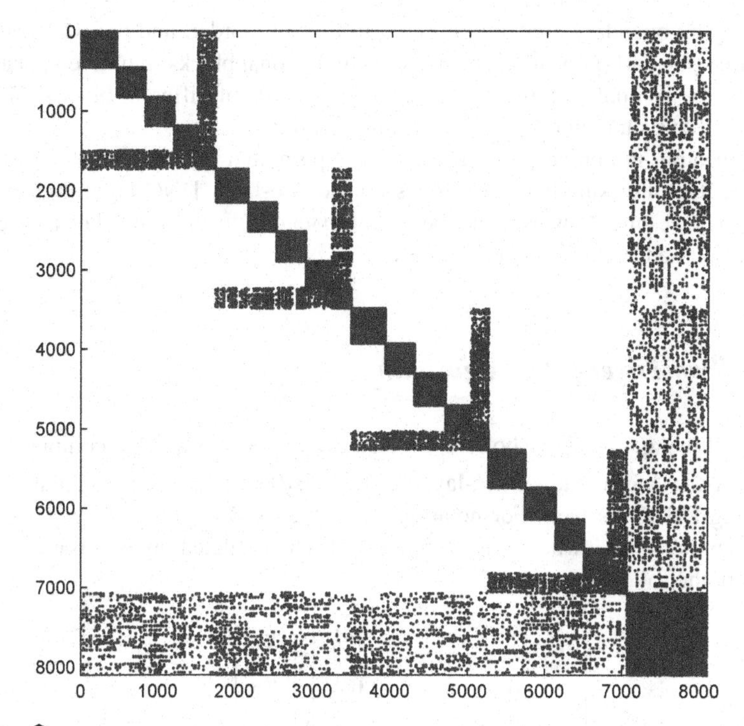

Fig. 4.4 $\widehat{\mathbf{A}}$ in a two-layer nested DBBD form after the second time RRH labelling, with $r_1 = 15\,\text{km}$, $r_2 = 8\,\text{km}$, where $N = 8100$, $r = 30\,\text{km}$, $d_0 = 500\,\text{m}$

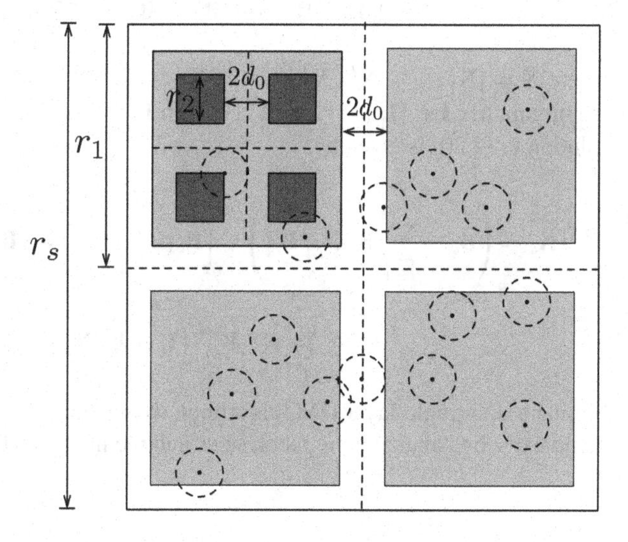

Fig. 4.5 Geographical RRH grouping in C-RAN

the light-grey boundary area are clustered to the sub-border, and the RRHs at the center area (the dark-grey area) are clustered to diagonal blocks. Intuitively, one can minimize the computation time by balancing the sizes of different blocks. This is the focus of our study in the remainder of this section.

By repeating the process, $\widehat{\mathbf{A}}$ can be further permuted into a multi-layer nested DBBD matrix. For simplicity, we focus on the two-layer DNC algorithm in this section. The results, however, can be easily extended to the multi-layer case, as briefly discussed at the end of the section.

4.3.1 Two-Layer DNC Algorithm

In the following, we show how $\widehat{\mathbf{A}}_{i,i}^{-1}$ can be computed with low computational complexity, if $\widehat{\mathbf{A}}$ is already a two-layer nested DBBD matrix with diagonal blocks $\widehat{\mathbf{A}}_{i,i}$ being DBBD as well. For notational brevity, let $\mathbf{B} \in \mathbb{C}^{L\times L}$ be an arbitrary diagonal block of $\widehat{\mathbf{A}}_{i,i}$, where L is the block size. Then, calculating $\mathbf{B}^{-1}$ is equivalent to solving the following system:

$$\begin{bmatrix} \mathbf{B}_{1,1} & & & & \mathbf{B}_{c1}^{\mathrm{H}} \\ & \mathbf{B}_{2,2} & & & \mathbf{B}_{c2}^{\mathrm{H}} \\ & & \cdots & & \vdots \\ & & & \mathbf{B}_{m,m} & \mathbf{B}_{cm}^{\mathrm{H}} \\ \mathbf{B}_{c1} & \mathbf{B}_{c2} & \cdots & \mathbf{B}_{cm} & \mathbf{B}_{c} \end{bmatrix} \begin{bmatrix} \mathbf{X}_1 \\ \mathbf{X}_2 \\ \vdots \\ \mathbf{X}_m \\ \mathbf{X}_c \end{bmatrix} = \mathbf{I}, \tag{4.20}$$

where $\mathbf{X} = [\mathbf{X}_1^T, \cdots, \mathbf{X}_m^T, \mathbf{X}_c^T]^T$, with $\mathbf{X}_i \in \mathbb{C}^{n_i\times L}$, and $\mathbf{X}_c \in \mathbb{C}^{n_c\times L}$. Partition the identity matrix $\mathbf{I}$ as $[\mathbf{I}_1^T, \cdots, \mathbf{I}_m^T, \mathbf{I}_c^T]^T$, with $\mathbf{I}_i \in \mathbb{C}^{n_i\times L}$, and $\mathbf{I}_c \in \mathbb{C}^{n_c\times L}$. Then, the solution to (4.20) is given by

$$\mathbf{X}_c = \left(\mathbf{B}_c - \sum_{i=1}^{i=m} \mathbf{B}_{ci}\mathbf{B}_{i,i}^{-1}\mathbf{B}_{ci}^{\mathrm{H}}\right)^{-1} \left[\mathbf{B}_{c1}\mathbf{B}_{1,1}^{-1}, \cdots, \mathbf{B}_{c1}\mathbf{B}_{m,m}^{-1}, \mathbf{I}\right], \tag{4.21}$$

$$\mathbf{X}_i = \mathbf{B}_{i,i}^{-1}\left(\mathbf{I}_i - \mathbf{B}_{ci}^{\mathrm{H}}\mathbf{X}_c\right). \tag{4.22}$$

Similar to the single-layer DNC algorithm, we will derive the optimal computational complexity by balancing the block sizes and the number of blocks.

4.3.2 *Optimizing the Computational Complexity*

The computational complexity of each step in calculating $\mathbf{B}^{-1}$ is listed in Table 4.2. In Table 4.2, $N_{d,t}$ and $N_{b,t}$ are the average diagonal block size and the cut-node block size in the tth layer, respectively. m_2 is the average number of diagonal blocks in $\mathbf{B}$. $\xi_2 \ll N_{d,1}$ is the average number of non-zero entries per row in $\mathbf{B}$. Following the single-layer DNC algorithm, we can adjust the side length r_t in the RRH labelling algorithm to optimize the computational complexity. Denote the block-size ratio of a diagonal block and the cut-node block in the tth layer by N^{z_t}, i.e., $N^{z_t} = \frac{N_{d,t}}{N_{b,t}}$. By adjusting r_t, N^{z_t} goes from $N_{d,t-1}^{-1}$ to $N_{d,t-1}$. Then the upper bounds of the diagonal block size and the cut-node block size in each layer is given below.

Lemma 4.2 *In large square C-RANs, given the block size ratio N^{z_t} in the tth layer, the side length r_t is the solution of equation $(r_t - 2d_0)^2 = 4(r_t - d_0)d_0\frac{r_{t-1}^2}{r_t^2}N^{z_t}$. Moreover,*

$$N_{d,t} = \left(4d_0\beta_N^{\frac{1}{2}}N_{d,t-1}N^{z_t}\right)^{\frac{2}{3}} + o(N_{d,t-1}^{\frac{2}{3}}N^{\frac{2}{3}z_t}), \tag{4.23}$$

$$N_{b,t} = \left(4d_0\beta_N^{\frac{1}{2}}N_{d,t-1}N^{-\frac{z_t}{2}}\right)^{\frac{2}{3}} + o(N_{d,t-1}^{\frac{2}{3}}N^{-\frac{1}{3}z_t}), \tag{4.24}$$

$$\lim_{r_s\to\infty} m_t = \left((4d_0)^{-2}\beta_N^{-1}N_{d,t-1}N^{-2z_t}\right)^{\frac{1}{3}}. \tag{4.25}$$

Table 4.2 Computational complexity in evaluating (4.21) and (4.22)

Step	Operation	Complexity/operation	Total number of operations
1.1	$\mathbf{B}_{i,i}^{-1}$	$O(N_{d,2}^3)$	m_2
1.2	$\mathbf{B}_{ci}\mathbf{B}_{i,i}^{-1}\mathbf{B}_{ci}^{\mathrm{H}}$ $\mathbf{B}_{ci}\mathbf{B}_{i,i}^{-1}\mathbf{I}_i$	$O(\frac{\xi_2}{m_2}N_{b,2}N_{d,2})$	m_2
1.3	$\mathbf{B}_c - \sum_{i=1}^{i=m_2}\mathbf{B}_{ci}\mathbf{B}_{i,i}^{-1}\mathbf{B}_{ci}^{\mathrm{H}}$ $\mathbf{I}_c - \sum_{i=1}^{i=m_2}\mathbf{B}_{ci}\mathbf{B}_{i,i}^{-1}\mathbf{I}_i$	$O(m_2N_{b,2}^2)$	1
1.4	$\left(\mathbf{B}_c - \sum_{i=1}^{i=m_2}\mathbf{B}_{ci}\mathbf{B}_{i,i}^{-1}\mathbf{B}_{ci}^{\mathrm{H}}\right)^{-1}$ $\times\left(\mathbf{I}_c - \sum_{i=1}^{i=m_2}\mathbf{B}_{ci}\mathbf{B}_{i,i}^{-1}\mathbf{I}_i\right)$	$O(N_{b,2}^2N_{d,1})$	1
1.5	$\mathbf{I}_i - \mathbf{B}_{ci}^{\mathrm{H}}\mathbf{X}_c$	$O(\frac{\xi_2}{m_2}N_{b,2}N_{d,1})$	m_2
1.6	$\mathbf{B}_{i,i}^{-1}(\mathbf{I}_i - \mathbf{B}_{ci}^{\mathrm{H}}\mathbf{X}_c)$	$O(N_{d,2}^2N_{d,1})$	m_2

From Tables 4.1 and 4.2, we note that the computational complexity of the two-layer DNC algorithm is $O(((N_{d,2}^3 + N_{d,2}^2 N_{d,1})m_2 + N_{b,2}^2 N_{d,1} + N_{d,1}^2)m_1 + N_{b,1}^3)$. Then, based on Lemma 4.2, we obtain the optimal computational complexity in Proposition 4.3.

Proposition 4.3 *The lowest computational complexity of the two-layer DNC algorithm is $O(N^2)$ with the optimal $z_1 = 0$ and $z_2 = -\frac{1}{6}$.*

We see that the computational complexity of the two-layer DNC algorithm is lower than that of the single-layer case, i.e., $O(N^{\frac{15}{7}})$. In Sect. 4.3.3, we will show that when parallel computing is deployed, the computation time of the two-layer DNC algorithm is also lower than that of the single-layer case, i.e., $O(N^2)$.

4.3.3 Parallel Computing

Similar to the single-layer DNC algorithm, parallel computing can be adopted in the multi-layer case. A nested parallel computing architecture is illustrated in Fig. 4.6. We first calculate $\mathbf{B}^{-1}$ in a parallel manner since every diagonal block in $\widehat{\mathbf{A}}$ is in a DBBD form. We use Table 4.2 to illustrate the idea of calculating $\mathbf{B}^{-1}$ in parallel. Steps 1.1 and 1.2 in Table 4.2 are first carried out in the level-3 processing units. The results are fed back into the level-2 processing units, which are responsible for performing steps 1.3 and 1.4. Then, steps 1.5 and 1.6 are carried out in the level-3 parallel processing units. Afterwards, the level-2 processing units calculate matrices $\widehat{\mathbf{A}}_{ci}\widehat{\mathbf{A}}_{i,i}^{-1}\widehat{\mathbf{A}}_{ci}^{\mathrm{H}}$ and vectors $\widehat{\mathbf{A}}_{ci}\widehat{\mathbf{A}}_{i,i}^{-1}\mathbf{y}_i$ for further processing. Then, similar to the Single-layer case, we obtain the minimum computation time for the two-layer DNC algorithm as given in Proposition 4.4.

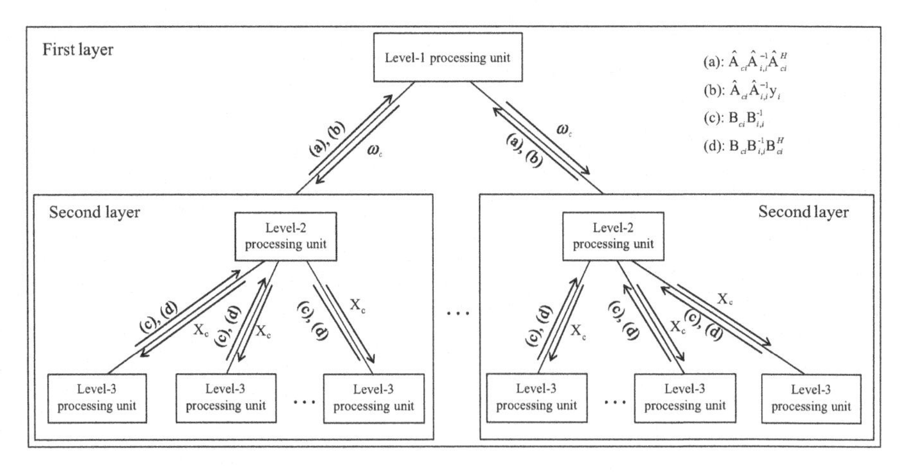

Fig. 4.6 Nested architecture of parallel computing of the two-layer DNC algorithm

Proposition 4.4 *Consider a C-RAN which has processing units with equal processing power. The optimal computation time, $O(N^{\frac{42}{23}})$, is achieved with $z_1 = \frac{4}{23}$ and $z_2 = 0$.*

As shown in Propositions 4.3 and 4.4, both the computational complexity and the computation time of the two-layer DNC algorithm is lower than that of the single-layer case. The computational complexity and computation time can be further reduced when more layers are introduced. We omit the details of implementation here due to space limitation.

4.4 Numerical Results

In this section, we compare the performance of the proposed DNC algorithm with that of a disjoint clustering algorithm. Unless stated otherwise, we assume that the minimum distance between RRHs and users is 1 m, the path loss exponent is 3.7, and the average transmit SNR at the user side is $\frac{P}{N_0} = 80\,\text{dB}$. As shown in Fig. 4.7, the disjoint clustering algorithm divides the whole network into disjoint square clusters with side length d_c. To reduce the channel estimation overhead, channel sparsification is also applied in the disjoint clustering algorithm. When the side length d_c is fixed, the computational complexity of disjoint clustering is linear with the network size. However, as shown in later simulations, the performance loss is noticeable. In Fig. 4.8a, we plot the SINR ratio against the distance threshold, where $N = 500$, $k = 500$, and the network area is a square with side length $r_s = 7\,\text{km}$. The SINR ratio of the DNC algorithm is plotted in blue, and that of the disjoint clustering algorithm is plotted in red. We see that the gap between the DNC algorithm and the disjoint clustering algorithm is very large. For example, there is about 10% SINR loss for disjoint clustering when the cluster size $d_c = 700\,\text{m}$. We then illustrate the

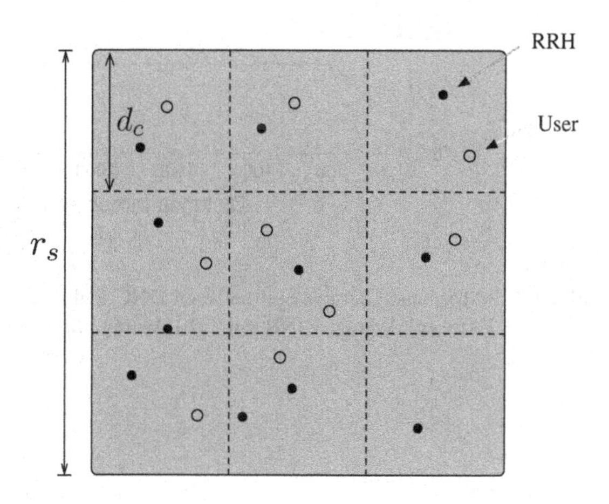

Fig. 4.7 Disjoint clustering in C-RAN

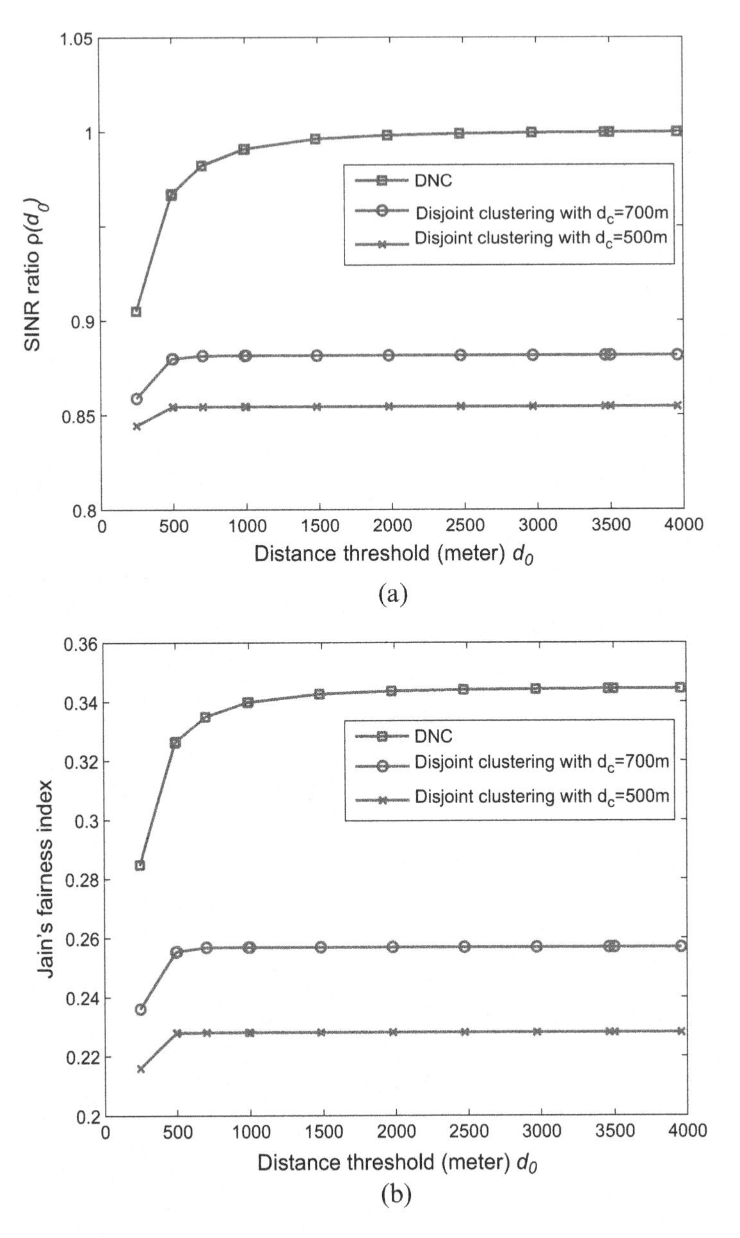

Fig. 4.8 Performance comparison between DNC and disjoint clustering when $N = 500$, $K = 500$, $r_s = 7\,\text{km}$. (**a**) Average SINR ratio. (**b**) Average Jain's fairness index

fairness loss caused by disjoint clustering in Fig. 4.8b. The fairness is quantified by the Jain's fairness index defined as

$$J(R_1, R_2, \cdots, R_K) = \frac{(\sum_{k=1}^{K} R_k)^2}{K \times \sum_{i=1}^{K} R_k^2}, \tag{4.26}$$

where $R_k = \log(1 + \text{SINR}_k)$ is the data rate of user k. The Jain's fairness index ranges from $\frac{1}{K}$ (the worst case) to 1 (the best case), and achieves maximum when all users have an equal data rate. Due to the randomness of the users' locations, the fairness index of the DNC algorithm is around 0.34 when the distance threshold is sufficiently large, say 700 m. We see that the fairness index of the disjoint clustering algorithm is much smaller than that of the DNC algorithm. This is because in the disjoint cluster algorithm, users in cluster edges always suffer from severe inter-cluster interference. To summarize, the DNC algorithm has a better performance than disjoint clustering in both average SINR and fairness.

4.5 Conclusions

In this chapter, we proposed the DNC algorithm to significantly enhance the scalability of the uplink signal processing in C-RANs. In the proposed algorithm, RRHs are dynamically clustered, i.e., each RRH is assigned either to one of the disjoint central clusters or to the boundary cluster based on their locations. The boundary cluster captures the interaction between central clusters, which avoids the performance loss caused by conventional clustering. We showed that both the size and the number of the central clusters as well as the size of the boundary cluster can be easily adjusted to minimize computational complexity. In addition, as the operations in central clusters can be performed in parallel, we also discussed clustering strategies to achieve the minimum computation time when parallel computing is deployed. Therefore, the DNC algorithm is adaptive to various computing implementations.

References

1. C. Fan, Y. J. Zhang, and X. Yuan, "Dynamic nested clustering for parallel PHY-layer processing in cloud-RANs," *IEEE Transactions on Wireless Communications*, vol. 15, no. 3, pp. 1881–1894, 2016.
2. Y. Saad, *Iterative methods for sparse linear systems*. SIAM, 2003.
3. S. Akoum and R. W. Heath, "Interference coordination: Random clustering and adaptive limited feedback," *IEEE Transactions on Signal Processing*, vol. 61, no. 7, pp. 1822–1834, 2013.
4. N. Lee, D. Morales-Jimenez, A. Lozano, and R. W. Heath, "Spectral efficiency of dynamic coordinated beamforming: A stochastic geometry approach," *IEEE Transactions on Wireless Communications*, vol. 14, no. 1, pp. 230–241, 2015.

Chapter 5
Scalable Signal Detection: Randomized Gaussian Message Passing

In this chapter, we convert the signal detection in a C-RAN to an inference problem over a bipartite random geometric graph. By passing messages among neighboring nodes, message passing (a.k.a. belief propagation) provides an efficient way to solve the inference problem over a sparse graph. However, the traditional message-passing algorithm does not guarantee to converge, because the corresponding bipartite random geometric graph is locally dense and contains many short loops. As a major contribution of this chapter, we propose a randomized Gaussian message passing (RGMP) algorithm to improve the convergence. The proposed RGMP algorithm demonstrates significantly better convergence performance than the conventional message passing algorithms. In addition, we generalize the RGMP algorithm to a blockwise RGMP (B-RGMP) algorithm, which allows parallel implementation. The average computation time of B-RGMP remains constant when the network size increases. The material in this chapter is mainly based on [1].

5.1 Gaussian Message Passing with Channel Sparsification

In this section, we first model a C-RAN as a bipartite random geometric graph. Then, we apply the Gaussian message-passing algorithm proposed in [2] over bipartite random geometric graphs for signal detection.

5.1.1 Bipartite Random Geometric Graph

Recall that the channel sparsification approach in Chap. 2 sparsifies the channel matrix by discarding the entries of the channel matrix **H** based on the distances between RRHs and users, and thereby simplifies the signal detection in a C-RAN

Y.-J. A. Zhang et al., *Scalable Signal Processing in Cloud Radio Access Networks*, SpringerBriefs in Electrical and Computer Engineering,
https://doi.org/10.1007/978-3-030-15884-2_5

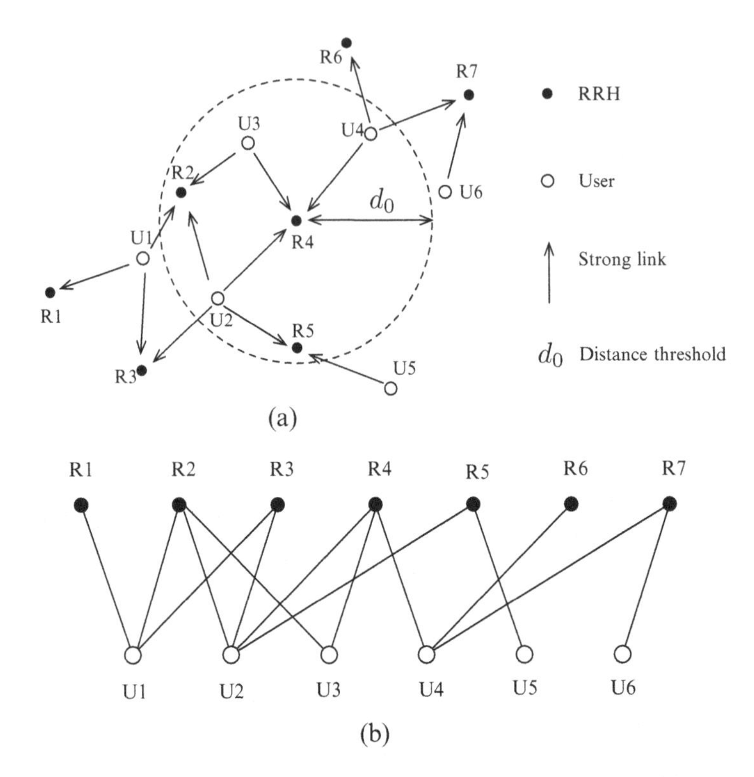

Fig. 5.1 Graphical representation of a C-RAN. (**a**) C-RAN architecture. (**b**) Bipartite random geometric graph

to an inference problem over a bipartite random geometric graph (see Fig. 5.1). In the bipartite random geometric graph, RRHs and users in a C-RAN are referred to RRH nodes and user nodes respectively, and edge connections exist only between RRH nodes and user nodes. More specifically, an RRH node is connected to a user node only if the distance between them falls below the threshold d_0, and the weight over such an edge is the channel coefficient from the corresponding user to the corresponding RRH.

Suppose that the entries in $\mathbf{x}$ follow an independent complex Gaussian distribution.[1] Then, $\mathbf{y}$ and $\mathbf{x}$ are jointly Gaussian, and therefore the MMSE detector is also the maximum *a posteriori* probability (MAP) detector that maximizes the *a posteriori* probability $p(\mathbf{x}|\mathbf{y})$ [4]. That is,

[1] If $\mathbf{x}$ does not follow a Gaussian distribution, the message-passing algorithm presented in this work gives an approximation of the linear MMSE estimation [3].

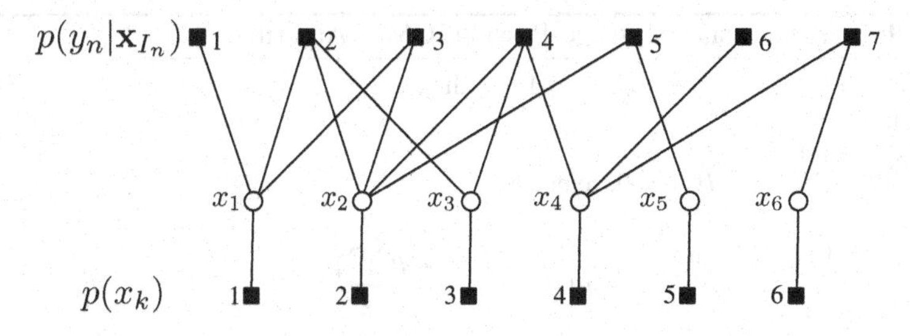

Fig. 5.2 A factor graph corresponding to the C-RAN in Fig. 5.1

$$\begin{aligned}\widehat{\mathbf{x}} &\approx P^{\frac{1}{2}}\widehat{\mathbf{H}}^{\mathrm{H}}\left(P\widehat{\mathbf{H}}\widehat{\mathbf{H}}^{\mathrm{H}}+\widehat{N}_0\mathbf{I}\right)^{-1}\mathbf{y} \\ &= \arg\max p(\mathbf{x}|\mathbf{y},\widehat{\mathbf{H}}),\end{aligned} \tag{5.1}$$

where for simplicity of notation, we assume that each user transmits with the same amount of power, say $P_1 = \cdots = P_K = P$. Based on (5.1), The probability density function $p(\mathbf{x}|\mathbf{y})$ can be factorized as

$$\begin{aligned}p(\mathbf{x}|\mathbf{y},\widehat{\mathbf{H}}) &\propto p(\mathbf{y}|\mathbf{x})p(\mathbf{x}) \\ &\approx p(y_1|\mathbf{x}_{\mathcal{I}_1})\cdots p(y_n|\mathbf{x}_{\mathcal{I}_n})\cdots p(y_N|\mathbf{x}_{\mathcal{I}_N}) \\ &\quad\times p(x_1)\cdots p(x_k)\cdots p(x_K),\end{aligned} \tag{5.2}$$

where $\mathbf{x}_{\mathcal{I}_n}$ contains all x_i with $i \in \mathcal{I}_n$ and $\mathcal{I}_n$ is the set of user indices with $d_{n,k} < d_0$.

We now transfer the bipartite random geometric graph to a factor graph with the factorization in (5.2). As illustrated in Fig. 5.2, a factor graph is also a bipartite graph comprising two types of nodes, namely, variable nodes (denoted by circles) and check nodes (denoted by squares), together with edges connecting these two types of nodes. The relation between the factorization (5.2) and its associated factor graph is as follows. A check node $p(y_n|\mathbf{x}_{\mathcal{I}_n})$ is connected to a variable node x_k by an edge when there is an edge connecting the n-th RRH node and the k-th user node in the corresponding random geometric graph (i.e., $d_{n,k} < d_0$), or equivalently, when the function $p(y_n|\mathbf{x}_{\mathcal{I}_n})$ takes x_k as input.

5.1.2 Gaussian Message Passing

We are now ready to introduce the Gaussian message-passing algorithm for signal detection. The algorithm will be implemented in the centralized data center. The messages, namely, the marginals of $\{x_k\}$ and $\{y_n\}$, are exchanged along the edges. In this chapter, both $\{x_k\}$ and $\{y_n\}$ are Gaussian distributed, and therefore the messages

Algorithm 5.1 Gaussian Message-Passing (GMP) Algorithm

1: Initial $t = 0$, $m^{(0)}_{x_k \to y_n} = 0$, $v^{(0)}_{x_k \to y_n} = 1$, for all k, n
2: **Repeat**
3: Set $t \Leftarrow t + 1$
4: For all n, k such that $\widehat{H}_{n,k} \neq 0$, compute

$$v^{(t)}_{y_n \to x_k} = \frac{1}{P|\widehat{H}_{n,k}|^2}\left(\widehat{N}_0 + P\sum_{j \neq k}|\widehat{H}_{n,j}|^2 v^{(t-1)}_{x_j \to y_n}\right) \tag{5.3}$$

$$m^{(t)}_{y_n \to x_k} = \frac{1}{P^{\frac{1}{2}}\widehat{H}_{n,k}}\left(y_n - P^{\frac{1}{2}}\sum_{j \neq k}\widehat{H}_{n,j} m^{(t-1)}_{x_j \to y_n}\right) \tag{5.4}$$

$$v^{(t)}_{x_k \to y_n} = \left(\sum_{\widehat{H}_{j,k} \neq 0, j \neq n} \frac{1}{v^{(t)}_{y_j \to x_k}} + 1\right)^{-1} \tag{5.5}$$

$$m^{(t)}_{x_k \to y_n} = v^{(t)}_{x_k \to y_n} \sum_{\widehat{H}_{j,k} \neq 0, j \neq n} \frac{m^{(t)}_{y_j \to x_k}}{v^{(t)}_{y_j \to x_k}} \tag{5.6}$$

5: **Until** the stopping criterion is satisfied
6: Compute

$$v_k = \left(\sum_{\widehat{H}_{n,k} \neq 0} \frac{1}{v^{(t)}_{y_n \to x_k}} + 1\right)^{-1} \tag{5.7}$$

$$\widehat{x}_k = v_k \sum_{\widehat{H}_{n,k} \neq 0} \frac{m^{(t)}_{y_n \to x_k}}{v^{(t)}_{y_n \to x_k}}. \tag{5.8}$$

are Gaussian probability density functions and can be completely characterised by mean and variance. Denote by $m^{(t)}_{y_n \to x_k}$ and $v^{(t)}_{y_n \to x_k}$ the mean and variance sent from check node $p(y_n|\mathbf{x}_{\mathcal{I}_n})$ to variable node x_k at iteration t, respectively, and denote by $m^{(t)}_{x_k \to y_n}$ and $v^{(t)}_{x_k \to y_n}$ the mean and variance sent from variable node x_k to check node $p(y_n|\mathbf{x}_{\mathcal{I}_n})$ at iteration t, respectively. The detailed steps of message passing are presented in Algorithm 5.1. We refer to this algorithm as Gaussian message passing (GMP), as all the messages involved are Gaussian marginals. Note that each RRH only serves users located in a circle with a constant radius d_0. Thus, the average number of messages to be exchanged and computed at each node does not scale with the network size. Therefore, the complexity per iteration of the GMP algorithm is linear in the number of RRHs and users.

In spite of its linear complexity per iteration, the GMP algorithm is not guaranteed to converge on the factor graphs induced by C-RANs. It is known that the GMP algorithm always converges to the optimal solution on a tree-type factor graph[2] [5]. It is also known that, if a factor graph is random and sparse enough, the corresponding message-passing algorithm converges asymptotically as the network size grows to infinity [6]. However, the factor graph for a bipartite random geometric graph induced from a C-RAN is locally dense and far from being a tree. This is due to the fact that every RRH needs to simultaneously serve multiple nearby users. For example, {R2, U1, R3, U2} in Fig. 5.1 form a loop[3] of length 4. Indeed, we observe in simulations that the GMP algorithm diverges in C-RAN with a non-trivial probability. Even worse, the probability of divergence grows with the network size, as illustrated later in Fig. 5.4. We focus on improving the convergence performance of GMP in the rest of the chapter.

Remark 5.1 The GMP algorithm for a C-RAN with channel sparsification can be simply extended to the case without channel sparsification by setting the distance threshold to infinity. However, this leads to an increase of the computational complexity per iteration. We see that in each iteration of Algorithm 5.1, messages need to be updated on every edge of the factor graph. From channel randomness, the entries of $\mathbf{H}$ are non-zero with probability one. Thus, in the factor graph without channel sparsification, every RRH check node $p(y_n|\mathbf{x})$ is connected to all variable nodes $\{x_k\}_{k=1}^{K}$. This implies that the total number of edges in the factor graph is NK, implying that the complexity of the GMP algorithm is $O(NK)$ per iteration, which is unaffordable for a large-scale C-RAN.

5.1.3 Related Work

A C-RAN is similar to a multiuser multiple-input multiple-output (MU-MIMO) system if the cooperative RRHs are regarded as multiple antennas of a single base station. Signal processing has been extensively studied in MU-MIMO systems. However, limited research has been focused on the scalability of signal processing complexity in MU-MIMO. Moreover, the distributed locations of RRHs make the distribution of the channel matrix in C-RAN distinct from that in MU-MIMO. As

[2] A tree-type graph is an undirected graph in which any two nodes are connected by exactly one path, where a path is a sequence of edges which connect a sequence of vertices without repetition.

[3] A loop in a graph is a path that starts and ends at the same node.

such, many existing results in MU-MIMO do not hold in C-RANs. For example, [7–9] proved the convergence of a message-passing-based detection algorithm for massive MU-MIMO system by exploiting the law of large numbers and the random matrix theory. However, in C-RANs, channel coefficients are dependent of each other since users and RRHs are geographically related to each other. Furthermore, even if we make the assumption that the channel coefficients are independent, the existing random matrix theory still does not apply, since the channel coefficients follow a truncated heavy-tailed distribution (which is not covered in the existing random matrix theory). A widely used method to improve the convergence of message passing is the damping technique [8–12]. With damping, an updated message is a weighted average of the message in the previous round of iteration and the message calculated by the original message updating rules. The weight in fact controls the trade-off between the convergence speed and the convergence probability. However, how to efficiently determine the value of the weight is still an open problem. It is also well-known that the schedule of message updating affects the convergence property of GMP [13]. Reference [13] analysed the average convergence speed of random serial update schedules for loop-free factor graphs. It has been proved that GMP with random serial schedules converges about twice as fast as the conventional GMP. The schedule analysed in [13] is randomly chosen and fixed in each realization instead of for each iteration. That is, the update schedule is the same for all iterations in [13]. As shown in our later simulations, the convergence of serial GMP heavily depends on the update order. With a randomly picked order, the serial GMP proposed in [13] does not ensure convergence. Another variant of message passing is approximate message passing (AMP). AMP was first proposed as a low-complexity iterative algorithm for compressed sensing [14]. Then, Rangan extended AMP to a general algorithm, named generalized approximate message passing (GAMP) [15]. However, in this chapter, we show by numerical simulations that for GAMP-based signal detection in C-RANs, the number of iterations needed for convergence is roughly linear in the network size. This translates to quadratic computational complexity in total, implying that GAMP is not scalable.

In [16], Shi et al. presented a two-stage approach to solve large-scale convex optimization problems for dense wireless cooperative networks, such as C-RANs. Matrix stuffing and alternating direction method of multipliers (ADMM) were used to speed up the computation. In addition, it was shown in [17] that the expected output of randomly permuted ADMM converges to the unique solution of the optimal linear detector. In this work we show that the ADMM algorithm converges much more slowly than the proposed RGMP algorithm when applied to large networks like C-RANs.

5.2 Randomized Gaussian Message Passing with Channel Sparsification

5.2.1 Randomized Gaussian Message Passing

In this section, we propose the RGMP algorithm to address the convergence issue of GMP. The main novelty of the RGMP algorithm is on the scheduling strategy for message updating. The conventional GMP algorithm employs synchronous message passing, i.e., messages are updated in parallel. As aforementioned, synchronous message passing does not work well in C-RANs due to local loops in the factor graph. It is well-known that serial message passing improves the convergence performance of GMP [13]. As shown in our later simulations, the convergence of GMP heavily depends on the update order. Nonetheless, there is no systematic way to derive a fixed update order that guarantees convergence. To address this issue, we propose the RGMP algorithm, which updates messages in a random order.

The RGMP algorithm is described as follows. At the t-th iteration, K random variables $\sigma_t(k)$ are generated from a continuous uniform distribution $(0, B)$. Then, the messages at the variable node x_k are updated at the time $\sigma_t(k)$. For example, when $K = 3$, $B = 1$, at the t-th iteration with $\sigma_t(1) = 0.623$, $\sigma_t(2) = 0.307$, $\sigma_t(3) = 0.890$, we first update all the messages on the edges connecting the variable node x_2 at time 0.307. Then, the messages on the edges connecting the variable node x_1 are updated at time 0.623. Finally, messages related to variable node x_3 are updated at time 0.890. The detailed RGMP algorithm is given in Algorithm 5.2.

Remark 5.2 In an ideal case when the updating and exchanging of messages does not incur any delay, each message can be updated with the up-to-date information. In this way, the RGMP algorithm updates the messages sequentially in a randomly permuted order. This is because the update time of messages is randomly generated continuous variables. With probability one, messages related to different variable nodes will not be updated at the same time. In reality, however, message updating and exchanging may cause a non-negligible delay, which means messages may not be updated with the latest information. If the update time interval B is much smaller than the delay, the RGMP algorithm is equivalent to the synchronous GMP. To make the RGMP algorithm different from synchronous GMP, the time interval B should be comparable to the overall delay in each iteration. Consequently, the computation time of RGMP will be significantly increased. Moreover, generating a large number of continuous random variables for creating the random schedule also introduces a non-negligible computational complexity. In Sect. 4.5, we will introduce a blockwise RGMP (B-RGMP) algorithm for parallel implementation. The total computation time of B-RGMP will not increase with the network size.

Algorithm 5.2 Randomized Gaussian Message-Passing (RGMP) Algorithm

Require: $\widehat{\mathbf{H}}$, $\mathbf{y}$

Ensure: $\widehat{x}_k$ for all k

1: initialize $t = 0$, $m_{x_k \to y_n}^{(0)} = 0$, $v_{x_k \to y_n}^{(0)} = 1$, for all k, n.
2: **Repeat**
3: Set $t \Leftarrow t + 1$.
4: Generate K random variables $\sigma_t(1), \cdots, \sigma_t(K)$ from a continuous uniform distribution on interval $(0, B)$.
5: For $i = 1, \cdots, K$, at time $\sigma_t(i)$, compute

$$v_{y_n \to x_i}^{(t)} = \frac{1}{P|\widehat{H}_{n,i}|^2}\left(\widehat{N}_0 + P \sum_{j:\sigma_t(j)<\sigma_t(i)} |\widehat{H}_{n,j}|^2 v_{x_{j)} \to y_n}^{(t)} + P \sum_{j \neq i:\sigma_t(j) \geq \sigma_t(i)} |\widehat{H}_{n,j}|^2 v_{x_j \to y_n}^{(t-1)}\right) \tag{5.9}$$

$$m_{y_n \to x_i}^{(t)} = \frac{1}{P^{\frac{1}{2}}\widehat{H}_{n,i}}\left(y_n - P^{\frac{1}{2}} \sum_{j:\sigma_t(j)<\sigma_t(i)} \widehat{H}_{n,j} m_{x_j \to y_n}^{(t)} - P^{\frac{1}{2}} \sum_{j \neq i:\sigma_t(j) \geq \sigma_t(i)} \widehat{H}_{n,j} m_{x_j \to y_n}^{(t-1)}\right) \tag{5.10}$$

$$v_{x_i \to y_n}^{(t)} = \left(\sum_{\widehat{H}_{j,i} \neq 0, j \neq n} \frac{1}{v_{y_j \to x_i}^{(t)}} + 1\right)^{-1} \tag{5.11}$$

$$m_{x_i \to y_n}^{(t)} = v_{x_i \to y_n}^{(t)} \sum_{\widehat{H}_{j,i} \neq 0, j \neq n} \frac{m_{y_j \to x_i}^{(t)}}{v_{y_j \to x_i}^{(t)}} \tag{5.12}$$

6: **Until** stopping criteria is satisfied
7: Compute

$$v_k = \left(\sum_{\widehat{H}_{n,k} \neq 0} \frac{1}{v_{y_n \to x_k}^{(t)}} + 1\right)^{-1} \tag{5.13}$$

$$\widehat{x}_k = v_k \sum_{\widehat{H}_{n,k} \neq 0} \frac{m_{y_n \to x_k}^{(t)}}{v_{y_n \to x_k}^{(t)}}. \tag{5.14}$$

5.2.2 *Numerical Examples*

In this subsection, we use a simple example to illustrate the difference between our proposed RGMP algorithm and synchronous/asynchronous GMP. Consider the following randomly generated channel matrix

$$\mathbf{H} = 10^{-5} \begin{bmatrix} -0.146 + 0.240i & -2.100 - 0.735i & -2.146 - 2.028i & 0.613 + 2.042i \\ 17.720 + 18.832i & 1.843 - 2.418i & 5.744 + 2.054i & 0.484 - 3.038i \\ 5.171 - 14.529i & 0.118 - 1.531i & -10.301 + 0.105i & 2.439 - 0.855i \\ -25.204 - 16.276i & 1.170 - 0.379i & 2.286 - 0.286i & 6.043 - 2.632i \end{bmatrix} \tag{5.15}$$

and let the transmit SNR (i.e., $\frac{P}{N_0}$) be 100 dB. The corresponding received signal $\mathbf{y}$ is

$$\mathbf{y} = \begin{bmatrix} 1.685 - 7.128i \\ -20.979 + 3.605i \\ -3.021 + 3.804i \\ 21.531 + 6.531i \end{bmatrix}. \tag{5.16}$$

For fair comparison, we do not conduct channel sparsification in this example. That is, the distance threshold is set to infinity. Figure 5.3 plots the relative error versus the number of iterations for the RGMP algorithm with $B = 1$ and the GMP algo-

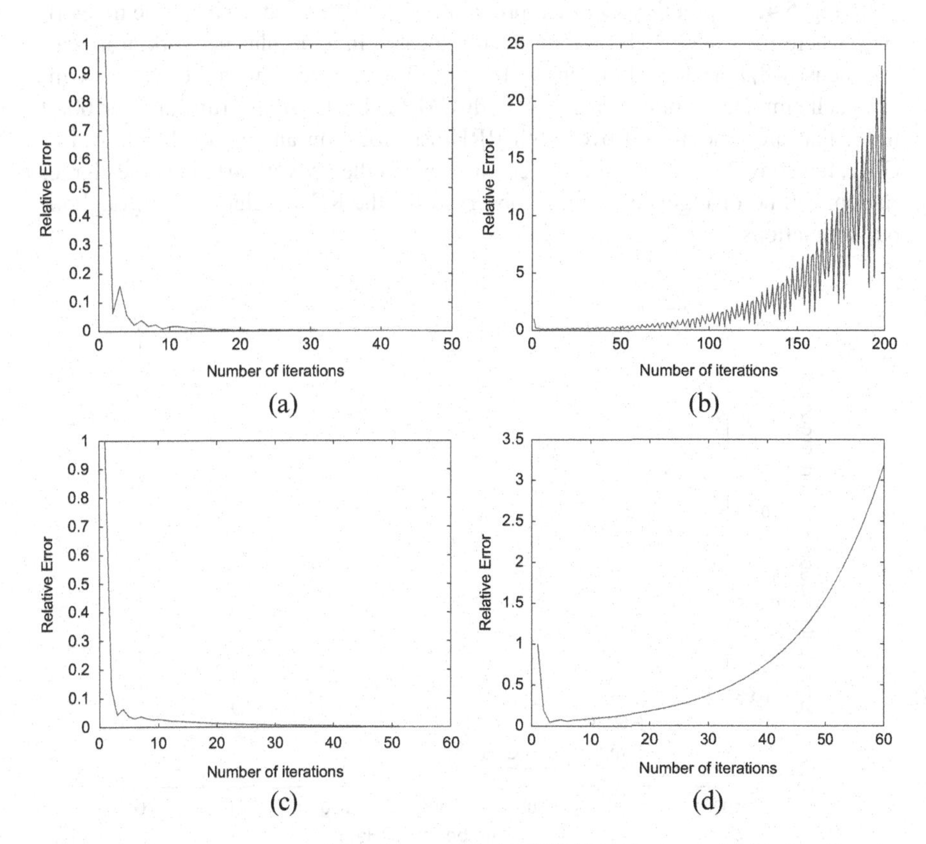

Fig. 5.3 Relative error vs number of iterations. (**a**) RGMP with $B = 1$. (**b**) Synchronous Gaussian message passing. (**c**) Asynchronous Gaussian message passing with schedule $\sigma = (0.198, 0.432, 0.909, 0.859)$. (**d**) Asynchronous Gaussian message passing with schedule $\sigma = (0.198, 0.432, 0.859, 0.909)$

rithm with different message updating strategies, i.e., synchronous update and asynchronous update with different fixed schedules, $\sigma = (0.198, 0.432, 0.909, 0.859)$ and $\sigma = (0.198, 0.432, 0.859, 0.909)$. The relative error is defined as $\frac{\|P\widehat{\mathbf{H}}^{\mathrm{H}}\widehat{\mathbf{H}}\mathbf{x}^{(t)} - P^{\frac{1}{2}}\widehat{\mathbf{H}}^{\mathrm{H}}\mathbf{y}\|}{\|P^{\frac{1}{2}}\widehat{\mathbf{H}}^{\mathrm{H}}\mathbf{y}\|}$, where $\mathbf{x}^{(t)}$ is the estimation of the transmitted signal after the t-th iteration. We see that the synchronous GMP algorithm and the asynchronous one with schedule $(0.198, 0.432, 0.859, 0.909)$ diverge, but the asynchronous GMP with schedule $(0.198, 0.432, 0.909, 0.859)$ and the proposed RGMP algorithm converge.

Remark 5.3 The examples in Fig. 5.3c, d show that convergence of asynchronous GMP heavily depends on the update schedule/order. Unfortunately, there is no systematic way to derive a fixed update order that guarantees convergence. In general, finding such an update order is difficult, especially in large networks. This issue is avoided in the proposed RGMP algorithm by randomizing the update schedule instead of fixing one. Indeed, the randomization significantly weakens the loopy effect of the graph, and thus convergence is almost ensured in RGMP.

In Fig. 5.4, we plot the empirical probability of convergence against the network size, where users and RRHs are uniformly located in a circular network area with user density $8/\text{km}^2$ and RRH density $10/\text{km}^2$. The distance threshold d_0 is 1000 m. For each simulated point in Fig. 5.4, both GMP and RGMP are run for over 6000 times that are randomized over both RRH/user location and channel fading. For GMP, the convergence probability decreases when the network size becomes large. In contrast, no divergence has been observed for the RGMP algorithm throughout our simulations.

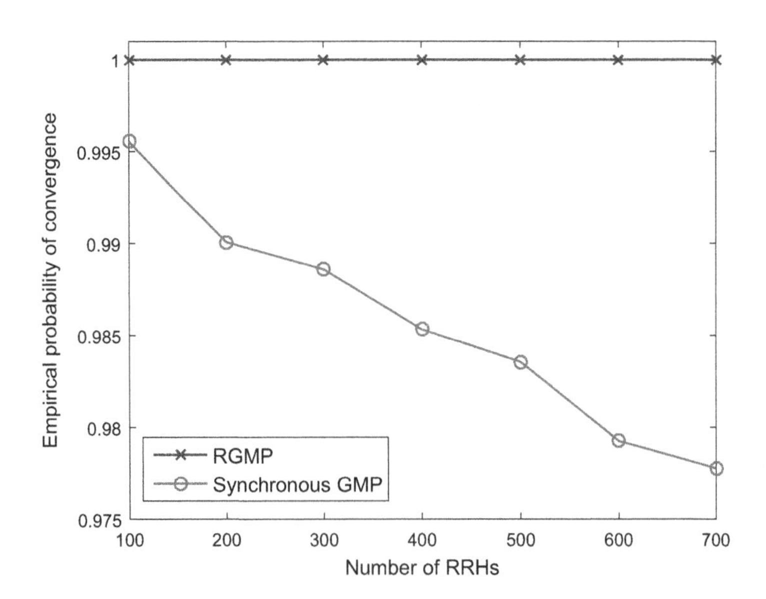

Fig. 5.4 Probability of convergence with $\beta_N = 10/\text{km}^2$, $\beta_K = 8/\text{km}^2$, and $P = 95\,\text{dB}$

5.3 Convergence Analysis

It is proven that the fixed point of GMP always provides the exact marginals (i.e., the solution of MMSE detection in this book), provided that the algorithm converges [3]. Thus, we only need to consider the convergence of the proposed algorithm, since the algorithm always gives the true solution of MMSE detection as long as it converges. In this section, we establish a necessary and sufficient condition for the expected convergence of the proposed RGMP algorithm. For self-containedness, we start with existing results on the analysis of the convergence condition for conventional GMP.

5.3.1 Convergence of GMP

The factor graph of a C-RAN contains loops with high probability. The convergence of GMP on a loopy factor graph has been previously studied in [18], with the main result summarized below.

From Algorithm 5.1, we see that the evolution of the variances $v_{y_n\to x_k}$ is independent of the means $m_{y_n\to x_k}$, $m_{x_k\to y_n}$ and the received signal $\mathbf{y}$. Substituting (5.5) into (5.3), we obtain

$$v^{(t)}_{y_n\to x_k} = \frac{\widehat{N}_0 + P\sum_{j\neq k}|\widehat{H}_{n,j}|^2\left(\sum_{\widehat{H}_{i,j}\neq 0, i\neq n}\frac{1}{v^{(t-1)}_{y_i\to x_j}}+1\right)^{-1}}{P|\widehat{H}_{n,k}|^2}. \tag{5.17}$$

Denote (5.17) in a vector form as

$$\mathbf{v}^{(t)} = f(\mathbf{v}^{(t-1)}), \tag{5.18}$$

where $f(\cdot)$ is the evolution function determined by (5.17), and $\mathbf{v}^{(t)}$ is a vector consisting of $v^{(t)}_{y_n\to x_k}$ for all n and k with $\widehat{H}_{n,k}\neq 0$. Note that $f(\cdot)$ is a standard function, the definition of which is given below.

Definition 5.1 A function $f(\mathbf{v})$ is standard if for all $\mathbf{v}\geq \mathbf{0}$ the following properties are satisfied.

- *Positivity:* $f(\mathbf{v}) > \mathbf{0}$.
- *Monotonicity:* If $\mathbf{v}\geq\mathbf{v}'$, then $f(\mathbf{v})\geq f(\mathbf{v}')$.
- *Scalability:* For all $\alpha>1$, $\alpha f(\mathbf{v}) > f(\alpha\mathbf{v})$.

Furthermore, we prove that the variances of GMP always converge to a unique fixed point in Lemma 5.1.[4]

[4] Lemma 5.1 was previously shown in Theorem 5.1 of [18], but the proof has been omitted in [18]. Here, we include the detailed proof of Lemma 1 for self-containedness.

Lemma 5.1 *In the GMP algorithm, if the initial point* $\mathbf{v}^{(0)} > \mathbf{0}$*, the sequence of* $\mathbf{v}^{(t)}$ *always converges to a fixed point of* $f(\cdot)$ *and the fixed point is unique.*

Proof As proven in Theorem 2 of [19], if a standard function has a fixed point and the initial point is positive, the algorithm always converges to a unique fixed point of the standard function. Thus, it suffices to show that $f(\cdot)$ has a fixed point. Since $\widehat{N}_0 > 0$, we suppose that $\mathbf{0} < \mathbf{v}^{(0)} < \frac{\widehat{N}_0}{P|\widehat{H}_{n,k}|^2} \cdot \mathbf{1}$ for all $\widehat{H}_{n,k} \neq 0$. Then, we can see that $\mathbf{v}^{(0)} < \mathbf{v}^{(1)}$. Consequently, we obtain $\mathbf{v}^{(0)} < \mathbf{v}^{(1)} < \cdots < \mathbf{v}^{(t-1)} < \mathbf{v}^{(t)}$. The sequence of variances is an increasing sequence. Moreover, the sequence is upper bounded by $c \cdot \mathbf{1}$, where c satisfies the following conditions

$$c \geq \frac{1}{P|\widehat{H}_{n,k}|^2}\left(\widehat{N}_0 + P\sum_{j\neq k}|\widehat{H}_{n,j}|^2\right), \forall \widehat{H}_{n,k} \neq 0. \tag{5.19}$$

Thus, the sequence of variances always converges to a limit point, and the limit point is a fixed point of $f(\mathbf{v})$. This concludes the proof.

We now consider the convergence of means. A vector of means, $\mathbf{m}^{(t)}$, is constructed with its $((k-1)N+n)$-th entry being

$$m^{(t)}_{(k-1)N+n} = \begin{cases} m^{(t)}_{y_n \to x_k}, & \widehat{H}_{n,k} \neq 0, \\ 0, & \text{otherwise}. \end{cases} \tag{5.20}$$

The recursion of the means is given by (5.4) and (5.6). As the variances always converge, the evolution of the means can be written as follows:

$$\mathbf{m}^{(t)} = \mathbf{\Omega}\mathbf{m}^{(t-1)} + \mathbf{z}, \tag{5.21}$$

where $\mathbf{z}$ is an $NK \times 1$ vector with its $((k-1)N+n)$-th entry being

$$z_{(k-1)N+n} = \begin{cases} \frac{y_n}{P^{\frac{1}{2}}\widehat{H}_{n,k}}, & \widehat{H}_{n,k} \neq 0, \\ 0, & \text{otherwise}, \end{cases} \tag{5.22}$$

and $\mathbf{\Omega}$ is an $NK \times NK$ matrix with the $((k-1)N+n, (j-1)N+i)$-th entry being

$$\Omega_{(k-1)N+n,(j-1)N+i} = \begin{cases} -\frac{\widehat{H}_{n,j} v^*_{x_j \to y_n}}{\widehat{H}_{n,k} v^*_{y_i \to x_j}}, & \widehat{H}_{n,k} \neq 0, \widehat{H}_{i,j} \neq 0, n \neq i, \text{and } j \neq k, \\ 0, & \text{otherwise}, \end{cases} \tag{5.23}$$

with $v^*_{x_j \to y_n} = (\sum_{\widehat{H}_{i,j} \neq 0, i \neq n} \frac{1}{v^*_{y_i \to x_j}} + 1)^{-1}$ and $v^*_{y_i \to x_j} = \lim_{t \to \infty} v^{(t)}_{y_i \to x_j}$. Then, a necessary and sufficient condition for the convergence of (5.21) is given in Theorem 5.2, [18]. That is, in Algorithm 5.1, the sequence of $\mathbf{m}^{(t)}$ converges to a unique fixed point if and only if the spectral radius $\rho(\boldsymbol{\Omega}) < 1$.

5.3.2 *Convergence of RGMP*

In this subsection, we first show that the message variances always converge in the RGMP algorithm. Then, we focus on the convergence condition of the means in RGMP.

Recall that the evolution function of the variances in (5.17) is a standard function with a unique fixed point. As proven in [19], if the evolution function of a synchronous algorithm is standard and feasible, then the corresponding asynchronous algorithm converges. Based on that, we obtain the following theorem.

Theorem 5.1 *In the RGMP algorithm, the sequence of* $v^{(t)}_{y_n \to x_k}$ *always converges to the same unique fixed point as in Algorithm 5.1 if the initial point* $\mathbf{v} > \mathbf{0}$.

With Theorem 5.1, it suffices to focus on the convergence condition of the means in the RGMP algorithm. Denote the update schedule at the t-th iteration as σ_t. Combining (5.10) and (5.12), we obtain the evolution of means $\mathbf{m}^{(t)}$ as

$$\mathbf{m}_k^{(t+1)} = \sum_{j:\sigma_t(j)<\sigma_t(k)} \boldsymbol{\Omega}_{k,j} \mathbf{m}_j^{(t+1)} + \sum_{j:\sigma_t(j)\geq\sigma_t(k)} \boldsymbol{\Omega}_{k,j} \mathbf{m}_j^{(t)} + \mathbf{z}_k, \tag{5.24}$$

where $\mathbf{m}_j^{(t)}$ is an $N \times 1$ subvector of $\mathbf{m}^{(t)}$ with the n-th entry being

$$m_j^{(t)}(n) = \begin{cases} m^{(t)}_{y_n \to x_j}, & \widehat{H}_{n,j} \neq 0, \\ 0, & \text{otherwise}, \end{cases} \tag{5.25}$$

and $\mathbf{z}_k$ is an $N \times 1$ subvector of $\mathbf{z}$ with the n-th entry being

$$z_k(n) = \begin{cases} \frac{y_n}{P^{\frac{1}{2}} \widehat{H}_{n,k}}, & \widehat{H}_{n,k} \neq 0, \\ 0, & \text{otherwise}. \end{cases} \tag{5.26}$$

$\boldsymbol{\Omega}_{k,j}$ is the $N \times N$ evolution matrix from user j to user k with the (n, m)-th entry being

$$\boldsymbol{\Omega}_{k,j}(n,m) = \begin{cases} -\frac{\widehat{H}_{n,j} v^*_{x_j \to y_n}}{\widehat{H}_{n,k} v^*_{y_m \to x_j}}, & \widehat{H}_{n,k} \neq 0, \text{ and } \widehat{H}_{n,j} \neq 0, \text{ and } n \neq m, \\ 0, & \text{otherwise.} \end{cases} \tag{5.27}$$

$\boldsymbol{\Omega}_{k,j}$ is the (k, j)-th submatrix of $\boldsymbol{\Omega}$. More specifically,

$$\boldsymbol{\Omega} = \begin{bmatrix} \mathbf{0} & \boldsymbol{\Omega}_{1,2} & \cdots & \boldsymbol{\Omega}_{1,K-1} & \boldsymbol{\Omega}_{1,K} \\ \boldsymbol{\Omega}_{2,1} & \mathbf{0} & \boldsymbol{\Omega}_{2,3} & \cdots & \boldsymbol{\Omega}_{2,K} \\ \vdots & \ddots & \ddots & \ddots & \vdots \\ \boldsymbol{\Omega}_{K-1,1} & \cdots & \boldsymbol{\Omega}_{K-1,K-2} & \mathbf{0} & \boldsymbol{\Omega}_{K-1,K} \\ \boldsymbol{\Omega}_{K,1} & \boldsymbol{\Omega}_{K,2} & \cdots & \boldsymbol{\Omega}_{K-1,K} & \mathbf{0} \end{bmatrix}. \tag{5.28}$$

We can further rewrite the Eq. (5.24) as

$$\mathbf{L}_{\sigma_t} \mathbf{m}^{(t+1)} = \mathbf{R}_{\sigma_t} \mathbf{m}^{(t)} + \mathbf{z}, \tag{5.29}$$

where $\mathbf{L}_{\sigma_t} = [\mathbf{L}_{\sigma_t}(k, j)]_{k,j} \in \mathbb{C}^{NK \times NK}$ with its (k, j)-th submatrix being

$$\mathbf{L}_{\sigma_t}(k, j) = \begin{cases} -\boldsymbol{\Omega}_{k,j}, & \sigma_t(k) > \sigma_t(j) \\ \mathbf{I}, & k = j \\ \mathbf{0}, & \text{otherwise,} \end{cases} \tag{5.30}$$

and $\mathbf{R}_{\sigma_t} = [\mathbf{R}_{\sigma_t}(k, j)]_{k,j} \in \mathbb{C}^{NK \times NK}$ with its (k, j)-th submatrix being

$$\mathbf{R}_{\sigma_t}(k, j) = \begin{cases} \boldsymbol{\Omega}_{k,j}, & \sigma_t(k) \leq \sigma_t(j) \\ \mathbf{0}, & \text{otherwise.} \end{cases} \tag{5.31}$$

Based on the definition of $\mathbf{L}_{\sigma_t}$, the determinant of $\mathbf{L}_{\sigma_t}$ is always 1 or -1. It implies that $\mathbf{L}_{\sigma_t}$ is nonsingular. Then, multiplying both sides of (5.29) by $\mathbf{L}_{\sigma_t}^{-1}$, we obtain

$$\mathbf{m}^{(t+1)} = \mathbf{L}_{\sigma_t}^{-1} \mathbf{R}_{\sigma_t} \mathbf{m}^{(t)} + \mathbf{L}_{\sigma_t}^{-1} \mathbf{z}. \tag{5.32}$$

Consequently, we obtain the following condition for the convergence of the RGMP algorithm.

Proposition 5.1 *For a given sequence of update schedules* $(\sigma_1, \sigma_2, \cdots, \sigma_t, \cdots)$, *the RGMP algorithm converges to the fixed point* $(\mathbf{I} - \boldsymbol{\Omega})^{-1}\mathbf{z}$ *if and only if*

$$\lim_{t \to \infty} \mathbf{L}_{\sigma_t}^{-1} \mathbf{R}_{\sigma_t} \cdots \mathbf{L}_{\sigma_1}^{-1} \mathbf{R}_{\sigma_1} = \mathbf{0}, \tag{5.33}$$

where σ_t *is the update schedule at the* t*-th iteration.*

Proof For an arbitrary update schedule σ, the fixed point of

$$\mathbf{m}^{(t+1)} = \mathbf{L}_\sigma^{-1}\mathbf{R}_\sigma\mathbf{m}^{(t)} + \mathbf{L}_\sigma^{-1}\mathbf{z} \tag{5.34}$$

is given by

$$\mathbf{m}^* = (\mathbf{I} - \mathbf{L}_\sigma^{-1}\mathbf{R}_\sigma)^{-1}\mathbf{L}_\sigma^{-1}\mathbf{z}. \tag{5.35}$$

Substituting $\mathbf{R}_\sigma = \boldsymbol{\Omega} - \mathbf{I} + \mathbf{L}_\sigma$ into (5.35), we obtain $\mathbf{m}^* = (\mathbf{I} - \boldsymbol{\Omega})^{-1}\mathbf{z}$. Clearly, $\mathbf{m}^*$ is independent of the choice of schedule σ. Define $\mathbf{e}^{(t)} = \mathbf{m}^{(t)} - (\mathbf{I} - \boldsymbol{\Omega})^{-1}\mathbf{z}$. Then, $\mathbf{e}^{(t)} = \mathbf{L}_{\sigma_t}^{-1}\mathbf{R}_{\sigma_t}\mathbf{e}^{(t-1)}$, for any iteration number t. By recursion, we obtain $\mathbf{e}^{(t+1)} = \mathbf{L}_{\sigma_t}^{-1}\mathbf{R}_{\sigma_t}\cdots\mathbf{L}_{\sigma_1}^{-1}\mathbf{R}_{\sigma_1}\mathbf{e}^{(0)}$. Therefore, $\mathbf{m}^{(t)} \to \mathbf{m}^*$ provided that $\mathbf{L}_{\sigma_t}^{-1}\mathbf{R}_{\sigma_t}\cdots\mathbf{L}_{\sigma_1}^{-1}\mathbf{R}_{\sigma_1} \to \mathbf{0}$ as $t \to \infty$. This concludes the proof.

Proposition 5.1 discusses the convergence condition for a given sequence of update schedules $(\sigma_1, \sigma_2, \cdots, \sigma_t, \cdots)$. To quantify the average performance of RGMP over random update schedules, we consider expected convergence in the following, where the expectation is taken over all possible schedules. Let the expected output be

$$\boldsymbol{\phi}^{(t)} = \mathrm{E}_{\xi_{t-1}}[\mathbf{m}^{(t)}], \tag{5.36}$$

where $\xi_t = (\sigma_1, \cdots, \sigma_t)$ is the set of the update schedules after iteration t. We are now ready to present a necessary and sufficient condition for the convergence of $\boldsymbol{\phi}^{(t)}$.

Theorem 5.2 *The expected output* $\boldsymbol{\phi}^{(t)} = \mathrm{E}_{\xi_{t-1}}[\mathbf{m}^{(t)}]$ *converges to the unique point* $(\mathbf{I} - \boldsymbol{\Omega})^{-1}\mathbf{z}$ *if and only if the spectral radius* $\rho(\boldsymbol{\Lambda}) < 1$, *where*

$$\boldsymbol{\Lambda} \triangleq \mathrm{E}_\sigma[\mathbf{L}_\sigma^{-1}\mathbf{R}_\sigma]. \tag{5.37}$$

Proof Denote $\mathbf{A} \triangleq \mathrm{E}_\sigma[\mathbf{L}_\sigma^{-1}]$. Based on the definition of ϕ^t, we obtain

$$\begin{aligned}
\boldsymbol{\phi}^{(t+1)} &= \mathrm{E}_{\xi_t}[\mathbf{L}_{\sigma_t}^{-1}\mathbf{R}_{\sigma_t}\mathbf{m}^{(t)} + \mathbf{L}_{\sigma_t}^{-1}\mathbf{z}] \\
&= \mathrm{E}_{\sigma_t}\left[\mathrm{E}_{\xi_{t-1}}\left[\mathbf{L}_{\sigma_t}^{-1}\mathbf{R}_{\sigma_t}\mathbf{m}^{(t)} + \mathbf{L}_{\sigma_t}^{-1}\mathbf{z}\right]\right] \\
&= \boldsymbol{\Lambda}\boldsymbol{\phi}^{(t)} + \mathbf{A}\mathbf{z}.
\end{aligned} \tag{5.38}$$

From Theorem 5.3 in [20], the sequence of $\boldsymbol{\phi}^{(t)}$ converges to the fixed point $(\mathbf{I} - \boldsymbol{\Lambda})^{-1}\mathbf{A}\mathbf{z}$ if and only if the spectral radius $\rho(\boldsymbol{\Lambda}) < 1$. Then, it suffices to show that $(\mathbf{I} - \boldsymbol{\Lambda})^{-1}\mathbf{A}\mathbf{z} = (\mathbf{I} - \boldsymbol{\Omega})^{-1}\mathbf{z}$.

Note that $\mathbf{\Omega} = \mathbf{R}_\sigma - \mathbf{L}_\sigma + \mathbf{I}$. Substituting $\mathbf{R}_\sigma = \mathbf{\Omega} + \mathbf{L}_\sigma - \mathbf{I}$ into (5.37), we obtain

$$\begin{aligned} \mathbf{\Lambda} = & \ \mathrm{E}_\sigma[\mathbf{L}_\sigma^{-1}(\mathbf{\Omega} + \mathbf{L}_\sigma - \mathbf{I})] \\ = & \ \mathrm{E}_\sigma[\mathbf{L}_\sigma^{-1}(\mathbf{\Omega} - \mathbf{I}) + \mathbf{I}] \\ = & \ \mathbf{A}(\mathbf{\Omega} - \mathbf{I}) + \mathbf{I}. \end{aligned} \tag{5.39}$$

Recall that $\rho(\mathbf{\Lambda}) < 1$, and thus $\mathbf{I}-\mathbf{\Lambda}$ is nonsingular. Together with $\mathbf{I}-\mathbf{\Lambda} = \mathbf{A}(\mathbf{I}-\mathbf{\Omega})$ from (5.39), we see that both $\mathbf{A}$ and $\mathbf{I} - \mathbf{\Omega}$ are nonsingular. Hence,

$$(\mathbf{I} - \mathbf{\Lambda})^{-1}\mathbf{A}\mathbf{z} = (\mathbf{A}(\mathbf{I} - \mathbf{\Omega}))^{-1}\mathbf{A}\mathbf{z} = (\mathbf{I} - \mathbf{\Omega})^{-1}\mathbf{z}, \tag{5.40}$$

which concludes the proof.

As illustrated later in Fig. 5.5, $\rho(\mathbf{\Lambda})$ is more likely to take small values than $\rho(\mathbf{\Omega})$, which implies that RGMP converges with a higher probability than GMP. Indeed, we have run over 10,000 times, and $\rho(\mathbf{\Lambda}) < 1$ for all cases.

5.4 Blockwise RGMP and Its Convergence Analysis

The proposed RGMP algorithm is conceptually simple, but may be cumbersome in implementation, for that the serial message updating schedule prohibits parallel computation. In this section, we generalize RGMP to the blockwise RGMP (B-RGMP) algorithm, which is suitable for parallel message updating. We show that the B-RGMP algorithm has better convergence behavior than synchronous GMP.

5.4.1 *Blockwise RGMP*

In the B-RGMP algorithm, each iteration is divided into M timeslots. A variable node randomly selects a timeslot for message updating in each iteration and the selection in different iterations are independent. The B-RGMP algorithm is given in Algorithm 5.3.

For a fixed M, the computation time per iteration of the B-RGMP algorithm does not scale with the network size. The reason is that at each timeslot, the B-RGMP algorithm can be implemented in parallel by assigning the message updating of different variable nodes to different processors. Recall that the computational complexity per variable node in each timeslot remains constant when the network size increases. Hence, with a constant number of timeslots, the average computation time of B-RGMP remains constant when the network size increases.

Algorithm 5.3 Blockwise Randomized Gaussian Message-Passing (B-RGMP) Algorithm

Require: $\widehat{\mathbf{H}}$, $\mathbf{y}$
Ensure: $\widehat{x}_k$ for all k
1: Initialize $t = 0$, $m^{(0)}_{x_k \to y_n} = 0$, $v^{(0)}_{x_k \to y_n} = 1$, for all k, n.
2: **Repeat**
3: Set $t \Leftarrow t + 1$.
4: Draw K numbers from $\{1, \cdots, M\}$ with replacement, and define the K numbers as $\sigma_t(1), \cdots, \sigma_t(K)$.
5: For $m = 1, \cdots, M$
6: For $k = 1, \cdots, K$, and $\sigma_t(k) = m$, $\widehat{H}_{n,k} \neq 0$, compute

$$v^{(t)}_{y_n \to x_k} = \frac{1}{P|\widehat{H}_{n,k}|^2}\left(\widehat{N}_0 + P \sum_{j:\sigma_t(j)<m} |\widehat{H}_{n,j}|^2 v^{(t)}_{x_j \to y_n} + P \sum_{j \neq k:\sigma_t(j) \geq m} |\widehat{H}_{n,j}|^2 v^{(t-1)}_{x_j \to y_n}\right) \tag{5.41}$$

$$m^{(t)}_{y_n \to x_k} = \frac{1}{P^{\frac{1}{2}} \widehat{H}_{n,k}}\left(y_n - P^{\frac{1}{2}} \sum_{j:\sigma_t(j)<m} \widehat{H}_{n,j} m^{(t)}_{x_j \to y_n} - P^{\frac{1}{2}} \sum_{j \neq k:\sigma_t(j) \geq m} \widehat{H}_{n,j} m^{(t-1)}_{x_j \to y_n}\right) \tag{5.42}$$

$$v^{(t)}_{x_k \to y_n} = \left(\sum_{\widehat{H}_{j,k} \neq 0, j \neq n} \frac{1}{v^{(t)}_{y_j \to x_k}} + 1\right)^{-1} \tag{5.43}$$

$$m^{(t)}_{x_k \to y_n} = v^{(t)}_{x_k \to y_n} \sum_{\widehat{H}_{j,k} \neq 0, j \neq n} \frac{m^{(t)}_{y_j \to x_k}}{v^{(t)}_{y_j \to x_k}} \tag{5.44}$$

7: **Until** stopping criteria is satisfied
8: Compute

$$v_k = \left(\sum_{\widehat{H}_{n,k} \neq 0} \frac{1}{v^{(t)}_{y_n \to x_k}} + 1\right)^{-1} \tag{5.45}$$

$$\widehat{x}_k = v_k \sum_{\widehat{H}_{n,k} \neq 0} \frac{m^{(t)}_{y_n \to x_k}}{v^{(t)}_{y_n \to x_k}}. \tag{5.46}$$

5.4.2 Convergence Analysis of B-RGMP

The convergence condition of the RGMP algorithm can be readily extended to that of the B-RGMP algorithm by changing the distribution of update schedule σ accordingly. Moreover, the B-RGMP algorithm allows us to derive a simple convergence condition for the special case with $M = 2$, where in each iteration, all messages are updated within two timeslots. We show that in the special case, if the GMP algorithm converges, the expected output $\boldsymbol{\phi}^{(t)}$ of B-RGMP always converges, as formally stated below.

Corollary 5.1 *When the number of timeslots $M = 2$, the expected output $\boldsymbol{\phi}^{(t)}$ of the B-RGMP algorithm converges to the unique point if and only if the spectral radius $\rho(\frac{3}{4}\boldsymbol{\Omega} + \frac{1}{4}\boldsymbol{\Omega}^2) < 1$.*

Proof From Theorem 5.2, it suffices to show that $\boldsymbol{\Lambda} = \frac{3}{4}\boldsymbol{\Omega} + \frac{1}{4}\boldsymbol{\Omega}^2$. Based on (5.31), when there are only two timeslots, we obtain $(\boldsymbol{\Omega} - \mathbf{R}_\sigma)^2 = \mathbf{0}$. Then,

$$\begin{aligned}
&\mathbf{L}_\sigma^{-1}\mathbf{R}_\sigma \\
=&(\mathbf{I} - (\boldsymbol{\Omega} - \mathbf{R}_\sigma))^{-1}\mathbf{R}_\sigma \\
=&(\mathbf{I} + \boldsymbol{\Omega} - \mathbf{R}_\sigma)\mathbf{R}_\sigma \\
=&\mathbf{R}_\sigma + (\boldsymbol{\Omega} - \mathbf{R}_\sigma)(\mathbf{R}_\sigma - \boldsymbol{\Omega} + \boldsymbol{\Omega}) \\
=&\mathbf{R}_\sigma - \mathbf{R}_\sigma\boldsymbol{\Omega} + \boldsymbol{\Omega}^2.
\end{aligned} \tag{5.47}$$

Since the probability of $\sigma_t(k) \leq \sigma_t(j)$ for arbitrary k and j is

$$\begin{aligned}
&P(\sigma_t(k) \leq \sigma_t(j)) \\
=&1 - P(\sigma_t(k) > \sigma_t(j)) \\
=&1 - \frac{2^{K-2}}{2^K} \\
=&\frac{3}{4},
\end{aligned} \tag{5.48}$$

the expectation of $\mathbf{R}_\sigma$ is $\mathrm{E}_\sigma[\mathbf{R}_\sigma] = \frac{3}{4}\boldsymbol{\Omega}$. Hence,

$$\boldsymbol{\Lambda} = \frac{3}{4}\boldsymbol{\Omega} + \frac{1}{4}\boldsymbol{\Omega}^2. \tag{5.49}$$

The B-RGMP algorithm with $M = 2$ is not trivial since the convergence is greatly improved compared with GMP. Denote by $\Lambda_1, \cdots, \Lambda_{NK}$ the eigenvalues of $\boldsymbol{\Omega}$. Then, the eigenvalues of $\frac{3}{4}\boldsymbol{\Omega} + \frac{1}{4}\boldsymbol{\Omega}^2$ are $\frac{3}{4}\Lambda_1 + \frac{1}{4}\Lambda_1^2, \cdots, \frac{3}{4}\Lambda_{NK} + \frac{1}{4}\Lambda_{NK}^2$. The spectral radius of $\boldsymbol{\Omega}$ and $\frac{3}{4}\boldsymbol{\Omega}+\frac{1}{4}\boldsymbol{\Omega}^2$ are $\max_i |\Lambda_i|$ and $\max_i |\frac{3}{4}\Lambda_i+\frac{1}{4}\Lambda_i^2|$ respectively. Thus, $\rho(\frac{3}{4}\boldsymbol{\Omega} + \frac{1}{4}\boldsymbol{\Omega}^2) < 1$ always holds if $\rho(\boldsymbol{\Omega}) < 1$. But the converse does not hold in general. Take the channel matrix $\mathbf{H}$ in (5.15) as an example. The spectral radius of the corresponding $\boldsymbol{\Omega}$ is $\rho(\boldsymbol{\Omega}) = 1.0287$, while $\rho(\frac{3}{4}\boldsymbol{\Omega}+\frac{1}{4}\boldsymbol{\Omega}^2) = 0.9203$. This means that the expected output of B-RGMP converges while GMP diverges. Therefore, the condition for the expected convergence of the B-RGMP algorithm is less stringent than that of the synchronous message passing.

5.5 Numerical Comparisons

In this section, we compare the performance of RGMP and B-RGMP with other existing algorithms. Unless specified otherwise, we assume that both users and RRHs are uniformly at random located in a circular area with user density $\beta_K = 8/\text{km}^2$ and RRH density $\beta_N = 10/\text{km}^2$. The path loss exponent is 3.7, and the average transmit SNR at the user side equals to 95 dB. That is $\frac{P}{N_0} = 95\,\text{dB}$. Moreover, the stopping criteria is $\delta^{(t)} < \delta$, where $\delta^{(t)}$ is the relative error after the t-th iteration, where one iteration means all the messages are updated once. In particular, $\delta^{(t)} = \frac{\|P\mathbf{H}^{\text{H}}\mathbf{H}\mathbf{x}^{(t)} - P^{\frac{1}{2}}\mathbf{H}^{\text{H}}\mathbf{y}\|}{\|P^{\frac{1}{2}}\mathbf{H}^{\text{H}}\mathbf{y}\|}$, with $\mathbf{x}^{(t)}$ being the estimated transmitted signal after t iteration.

5.5.1 *Comparison of Convergence*

In this subsection, we compare the convergence of RGMP with conventional GMP. From Lemma 5.1 and Theorem 5.1, both GMP and RGMP have guaranteed convergence for variances. However, they have different conditions to ensure the convergence of means: GMP requires $\rho(\boldsymbol{\Omega}) < 1$ while RGMP requires $\rho(\boldsymbol{\Lambda}) < 1$. Since both $\boldsymbol{\Omega}$ and $\boldsymbol{\Lambda}$ highly depend on the network geometry, it is difficult to theoretically compare these two convergence conditions. To shed light on the difference, we plot the cumulative distribution function (CDF) of the spectral radius of $\boldsymbol{\Omega}$ and $\boldsymbol{\Lambda}$ in Fig. 5.5. We assume that users and RRHs are randomly located in a circular network area with radius r. The user density is $\beta_K = 8/\text{km}^2$ and the RRH density is $\beta_N = 10/\text{km}^2$. We see that $\rho(\boldsymbol{\Lambda})$ is more likely to take small values than $\rho(\boldsymbol{\Omega})$. This implies that RGMP converges with a higher probability than GMP. Indeed, we have run over 10,000 times for each setting, and $\rho(\boldsymbol{\Lambda}) < 1$ for all the cases. We also plot the CDF of $\rho(\boldsymbol{\Lambda})$ for B-RGMP with $M = 2$ and $M = 10$. We see that when the number of timeslots M is moderately large (e.g., $M = 10$), the distributions of $\rho(\boldsymbol{\Lambda})$ for RGMP and B-RGMP are quite close to each other. This implies that the B-RGMP algorithm is a reasonable generalization of RGMP without degrading the convergence performance.

In Fig. 5.5, we also compare the convergence of RGMP with damping-based GMP [8, 21]. In the GMP algorithm with damping, a message from the check nodes is a weighted average between the old message and the new message. That is, the Eq. (5.4) of Algorithm 5.1 is replaced by the following equation

$$m_{y_n \to x_k}^{(t)} = \eta \frac{y_n - P^{\frac{1}{2}} \sum_{j \neq k} \widehat{H}_{n,j} m_{x_j \to y_n}^{(t-1)}}{P^{\frac{1}{2}} \widehat{H}_{n,k}} + (1 - \eta) m_{y_n \to x_k}^{(t-1)}, \tag{5.50}$$

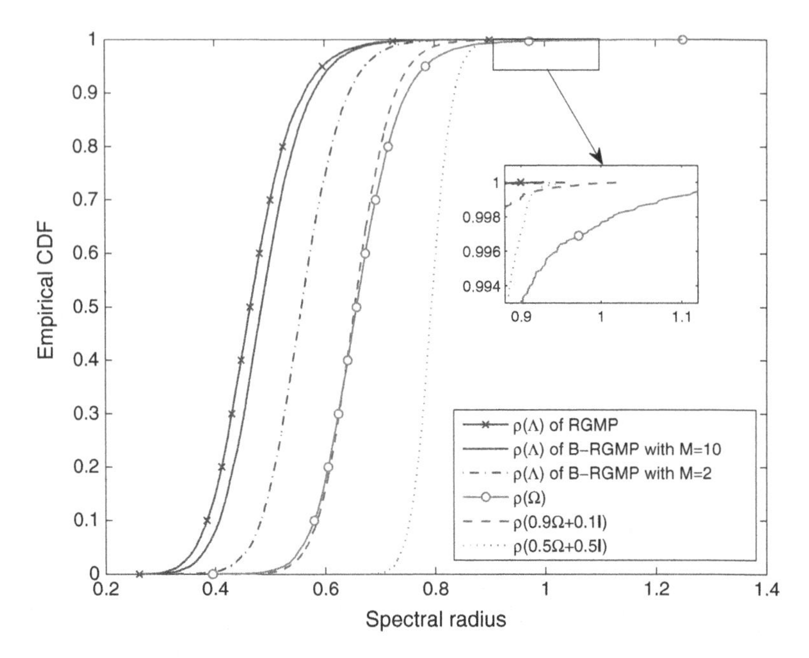

Fig. 5.5 Cumulative distribution function of the spectral radius with $N = 20$, $K = 15$, and $\frac{P}{N_0} = 95\,\text{dB}$

where η is the damping factor. Consequently, the convergence condition of GMP with damping now becomes $\rho(\eta\boldsymbol{\Omega} + (1 - \eta)\mathbf{I}) < 1$. As we can see from Fig. 5.5, when the damping factor is large (i.e., $\eta = 0.9$ in Fig. 5.5), the probability that the spectral radius exceeds 1 is non-zero. When the damping factor is small, the spectral radius is less likely to exceed 1. However, as shown in our later simulations, the convergence rate decreases when the damping factor decreases. There exists a trade-off between the convergence probability and the convergence rate of GMP with damping. How to efficiently determine the value of the damping factor is still an open problem.

5.5.2 *Comparison of Convergence Speed*

In this subsection, we compare the convergence speed of RGMP and B-RGMP with other algorithms including ADMM [17], GAMP [15], GMP with damping [21], and conjugate gradient (CG) [22]. For a fair comparison, the channel sparsification approach with distance threshold $d_0 = 1000\,\text{m}$ is adopted in all algorithms. In this way, all the algorithms have a linear per-iteration computational complexity with the network size. Thus, we only focus on the convergence speed of these algorithms.

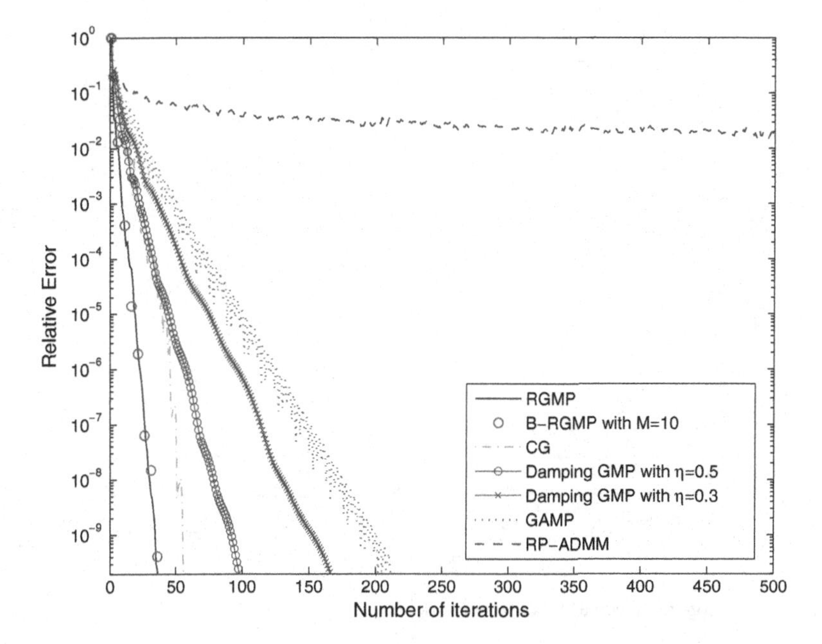

Fig. 5.6 Relative error vs number of iterations when the number of RRHs $N = 40$

In Fig. 5.6, the relative error $\delta^{(t)}$ is plotted against the number of iterations for $N = 40$ and $K = 32$. We see that RGMP, B-RGMP, GMP with damping, and CG converge relatively fast. For example, the relative error of RGMP reduces to 0.001 within 10 iterations. However, the performance of the ADMM algorithm is unsatisfactory. Over 500 iterations are needed for the ADMM algorithm to reduce the relative error to 0.02. In fact, from simulation results not presented here, ADMM requires over 5000 iterations on average to reduce the relative error to 0.001 for the network configuration in Fig. 5.6. Therefore, even though the computational complexity per iteration of ADMM is linear in the number of RRHs and the expected convergence is guaranteed [17], it is still impractical to adopt the ADMM algorithm in C-RAN due to the extremely slow convergence.

In Fig. 5.7, we plot the convergence speed of both RGMP and B-RGMP against the network size, where the convergence speed is measured by the critical number of iterations to achieve $\delta^{(t)} < 10^{-5}$. Due to the extremely slow convergence speed of ADMM as shown in Fig. 5.6, we ignore ADMM and only plot the convergence speed of GMP with damping (with $\delta^{(t)} < 10^{-5}$) and GAMP (with $\delta^{(t)} < 10^{-3}$) for comparison. We observe that the number of iterations needed by GAMP grows roughly linearly with the network size. In contrast, the convergence speeds of both RGMP, B-RGMP, and GMP with damping are constant with the network size. Note that the computational complexity per iteration of GAMP/GMP with damping/RGMP/B-RGMP is linear in the network size. Thus, the total computational complexity of both GMP with damping and the RGMP/B-RGMP

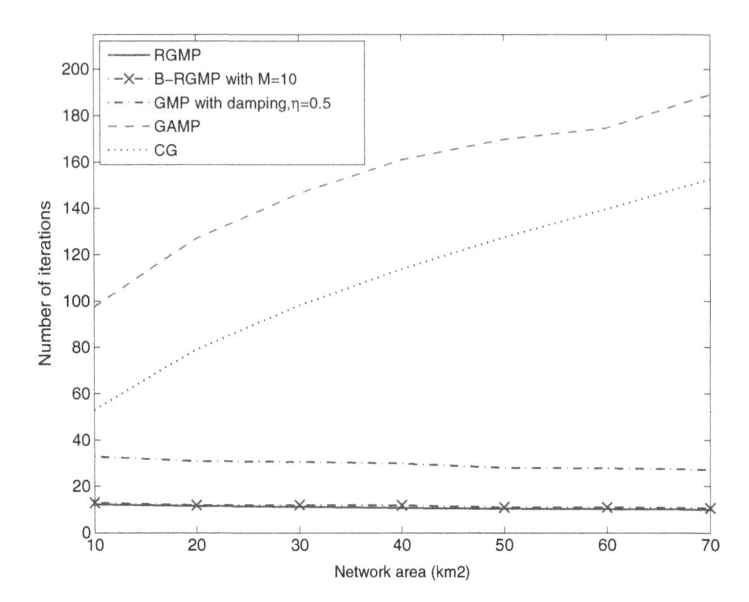

Fig. 5.7 Convergence speed against the network size

algorithm is linear in the network size, while that of GAMP grows quadratically with the network size. Moreover, with parallel implementation, the computation time of B-RGMP remains constant with the network size.

We emphasize that even though its performance looks not bad in simulation, GMP with damping has several drawbacks compared with the RGMP algorithm. For example, how to efficiently determine the value of the damping factor is still an open problem. In Fig. 5.6, we observe that the GMP with damping converges faster when the damping factor increases. Recall that the spectral radius is more likely to exceed 1 when the damping factor increases. Indeed, there exists a trade-off between the convergence probability and the convergence speed of GMP with damping. In previous works, the damping factor is usually determined through simulations [12]. Considering the large network size of C-RAN, empirically calculating the damping factor introduces unaffordable complexity cost. A recent work [21] derived a range of the damping factors, which guarantees the convergence of GMP. The range, however, is a function of the eigenvalues of $\mathbf{\Omega}$, which means choosing the damping factor based on [21] still requires prohibitively high computational complexity.

5.5.3 Comparison of Performance

In this subsection, we compare the performance of the RGMP algorithm with a disjoint clustering algorithm. The disjoint clustering algorithm divides the whole network into disjoint square clusters with area A_c, and do MMSE detection

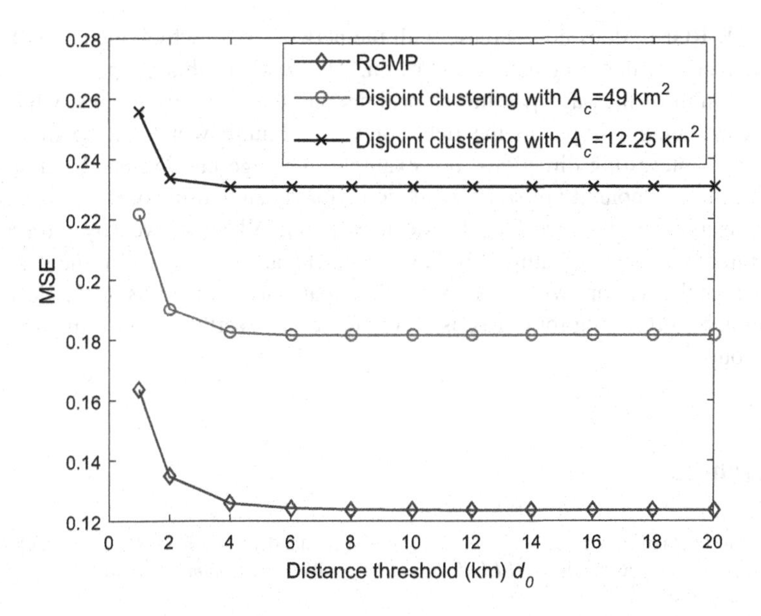

Fig. 5.8 SINR ratio vs the distance threshold d_0 when the network area is 200 km^2

independently in each disjoint cluster. Channel sparsification is also applied in the disjoint clustering algorithm. In Fig. 5.8, we plot the mean squared error (MSE) against the distance threshold, where MSE refers to $\mathrm{E}[|x_k - \hat{x}_k|^2]$. The network area is 200 km^2. The numbers of RRHs and users are 2000 and 1600, respectively. We see that the gap between the RGMP algorithm and the disjoint clustering algorithm is very large. For example, when the distance threshold d_0 is 4 km, the MSE of RGMP is less than 0.13, which is only half of the MSE of the disjoint clustering algorithm with cluster area 49 km^2.

5.6 Conclusions

In this chapter, we proposed RGMP and B-RGMP for scalable uplink signal detection in C-RANs. With channel sparsification, signal detection in a C-RAN was converted to an inference problem over a bipartite random geometric graph. A random message-update schedule was employed to address the convergence issue of GMP over a bipartite random geometric graph. We analysed the convergence condition of the proposed RGMP algorithm and showed that the convergence condition of RGMP is much less stringent than that of GMP. Numerical results demonstrated that RGMP exhibits much faster convergence than the existing algorithms, such as GAMP and ADMM. We further proposed the B-RGMP algorithm for parallel implementation. With a fixed number of timeslots, the total computation time of B-RGMP does not increase with the network size, which means B-RGMP is a perfectly scalable detection algorithm. The work in this chapter sheds light on the design of message-passing algorithms on general loopy graphs, which has been a challenging topic in the field for years. Future work can be done in a number of interesting directions. For example, message passing has been applied to reduce the complexity of signal detection with constellation constraints [11, 23]. The convergence of these algorithms can be potentially improved by introducing randomized message updating. Moreover, RGMP can be extended to the design of uplink signal detectors with limited fronthaul capacity, as well as to the design of downlink beamforming for C-RANs. These topics are worthy of our future research endeavour.

References

1. C. Fan, X. Yuan, and Y. J. Zhang, "Scalable uplink signal detection in C-RANs via randomized Gaussian message passing," *IEEE Transaction on Wireless Communications*, vol. 16, no. 8, pp. 5187–5200, 2017.
2. D. Bickson, D. Dolev, O. Shental, P. H. Siegel, and J. K. Wolf, "Gaussian belief propagation based multiuser detection," in *Proc. of IEEE International Symposium on Information Theory (ISIT)*, 2008, pp. 1878–1882.
3. Y. Weiss and W. T. Freeman, "Correctness of belief propagation in Gaussian graphical models of arbitrary topology," in *Advances in neural information processing systems*, 2000, pp. 673–679.
4. S. V. Vaseghi, *Advanced digital signal processing and noise reduction*. John Wiley & Sons, 2008.
5. F. R. Kschischang, B. J. Frey, and H.-A. Loeliger, "Factor graphs and the sum-product algorithm," *IEEE Transactions on Information Theory*, vol. 47, no. 2, pp. 498–519, 2001.
6. T. J. Richardson and R. L. Urbanke, "The capacity of low-density parity-check codes under message-passing decoding," *IEEE Transactions on Information Theory*, vol. 47, no. 2, pp. 599–618, 2001.
7. L. Liu, C. Yuen, Y. L. Guan, Y. Li, and Y. Su, "A low-complexity Gaussian message passing iterative detector for massive MU-MIMO systems," in *Proc. of IEEE International Conference on Information and Communications Security (ICICS)*, 2015, pp. 1–5.
8. L. Liu, C. Yuen, Y. L. Guan, Y. Li, and Y. Su, "Convergence analysis and assurance for Gaussian message passing iterative detector in massive MU-MIMO systems," *IEEE Transactions on Wireless Communications*, vol. 15, no. 9, pp. 6487–6501, 2016.
9. L. Liu, C. Yuen, Y. L. Guan, Y. Li, and C. Huang, "Gaussian message passing iterative detection for MIMO-NOMA systems with massive access," in *Proc. of IEEE Global Communications Conference (GLOBECOM)*, 2016, pp. 1–6.
10. I. Sohn, S. H. Lee, and J. G. Andrews, "Belief propagation for distributed downlink beamforming in cooperative MIMO cellular networks," *IEEE Transactions on Wireless Communications*, vol. 10, no. 12, pp. 4140–4149, 2011.
11. P. Som, T. Datta, A. Chockalingam, and B. S. Rajan, "Improved large-MIMO detection based on damped belief propagation," in *Proc. of IEEE Information Theory Workshop (ITW)*, 2010, pp. 1–5.
12. M. Moretti, A. Abrardo, and M. Belleschi, "On the convergence and optimality of reweighted message passing for channel assignment problems," *IEEE Signal Processing Letters*, vol. 21, no. 11, pp. 1428–1432, 2014.

13. J. Goldberger and H. Kfir, "Serial schedules for belief-propagation: Analysis of convergence time," *IEEE Transactions on Information Theory*, vol. 54, no. 3, pp. 1316–1319, 2008.
14. D. L. Donoho, A. Maleki, and A. Montanari, "Message-passing algorithms for compressed sensing," *Proceedings of the National Academy of Sciences*, vol. 106, no. 45, pp. 18 914–18 919, 2009.
15. S. Rangan, "Generalized approximate message passing for estimation with random linear mixing," in *IEEE International Symposium on Information Theory (ISIT)*, 2011, pp. 2168–2172.
16. Y. Shi, J. Zhang, B. O'Donoghue, and K. B. Letaief, "Large-scale convex optimization for dense wireless cooperative networks," *IEEE Transactions Signal Processing*, vol. 63, no. 18, pp. 4729–4743, 2015.
17. R. Sun, Z.-Q. Luo, and Y. Ye, "On the expected convergence of randomly permuted ADMM," *arXiv preprint arXiv:1503.06387*, 2015.
18. B. L. Ng, J. Evans, and S. Hanly, "Distributed downlink beamforming in cellular networks," in *Proc. of IEEE International Symposium on Information Theory (ISIT)*, 2007, pp. 6–10.
19. R. D. Yates, "A framework for uplink power control in cellular radio systems," *IEEE Journal on selected areas in communications*, vol. 13, no. 7, pp. 1341–1347, 1995.
20. O. Axelsson, *Iterative solution methods*. Cambridge university press, 1996.
21. Q. Su and Y.-C. Wu, "On convergence conditions of Gaussian belief propagation." *IEEE Transactions on Signal Processing*, vol. 63, no. 5, pp. 1144–1155, 2015.
22. R. Barrett, *et al.*, *Templates for the solution of linear systems: Building blocks for iterative methods*. SIAM, 1994.
23. S. Wu, *et al.*, "Low-complexity iterative detection for large-scale multiuser MIMO-OFDM systems using approximate message passing," *IEEE Journal of Selected Topics in Signal Processing*, vol. 8, no. 5, pp. 902–915, 2014.

Chapter 6
Conclusions and Future Work

6.1 Conclusions

Featuring centralized baseband processing, cooperative radio, and real-time cloud infrastructure, C-RAN has great potential to be a predominant wireless cellular architecture in next-generation wireless systems. Aware of the prohibitively high cost caused by the high RRH density, this book focuses on designing scalable signal processing algorithms for C-RANs by exploiting the near-sparsity of channel matrices, where "scalable" means that both the overhead and computational complexity does not grow significantly with the network size. The main results of the book are summarized as follows.

In Chap. 2, we investigated the effect of channel sparsification on the system performance. In particular, we proposed a threshold-based channel matrix sparsification method, and derived a closed-form expression describing the relationship between the threshold and the SINR loss due to channel sparsification. It was shown that a vast majority of the channel coefficients can be ignored with a small percentage of SINR loss. According to our simulations, by compromising only 5% SINR loss, the CSI acquisition of each RRH can be reduced from all users to a small number of closest users.

In Chap. 3, based on the channel sparsification approach, we considered the design of training sequences for time-multiplexing channel training in C-RANs. The training design problem was formulated so as to find the minimum length of training sequences that preserve local orthogonality, and a training design scheme based on graph coloring was proposed. The training length is $O(\ln K)$ almost surely as $K \to \infty$, where K is the number of users. Hence, the proposed training design can be applied to a large-size C-RAN satisfying local orthogonality at the cost of an acceptable training length.

Y.-J. A. Zhang et al., *Scalable Signal Processing in Cloud Radio Access Networks*, SpringerBriefs in Electrical and Computer Engineering,
https://doi.org/10.1007/978-3-030-15884-2_6

Chapters 4 and 5 proposed two low-complexity algorithms for signal detection based on channel sparsification. In Chap. 4, we proposed the DNC algorithm to significantly enhance the scalability of the uplink signal processing in C-RANs. In the proposed algorithm, RRHs are dynamically clustered, i.e., each RRH is assigned either to one of the disjoint central clusters or to the boundary cluster based on their locations. The boundary cluster captures the interaction between central clusters, which avoids the performance loss caused by conventional clustering. We showed that both the size and the number of the central clusters as well as the size of the boundary cluster can be easily adjusted to minimize computational complexity. In addition, as the operations in central clusters can be performed in parallel, we also discussed clustering strategies to achieve the minimum computation time when parallel computing is deployed. Therefore, the DNC algorithm is adaptive to various computing implementations. In Chap. 5, we proposed RGMP and B-RGMP for scalable uplink signal detection. With channel sparsification, signal detection in a C-RAN was converted to an inference problem over a bipartite random geometric graph. A random message-update schedule was employed to address the convergence issue of GMP over a bipartite random geometric graph. We analysed the convergence condition of the proposed RGMP algorithm and showed that the convergence condition of RGMP is much less stringent than that of GMP. Numerical results demonstrated that RGMP exhibits much faster convergence than the existing algorithms, such as GAMP and ADMM. We further proposed the B-RGMP algorithm for parallel implementation. With a fixed number of timeslots, the total computation time of B-RGMP does not increase with the network size, which means B-RGMP is a perfectly scalable detection algorithm.

In summary, we studied the scalability issues of signal processing in C-RANs. Our results established a tradeoff between the complexity and system performance through threshold-based channel sparsification.

6.2 Future Work

The work presented in this book offers many possibilities for future extensions. In particular, the following topics are of interest:

1. **Downlink Beamforming:**
   In this book, we only considered uplink signal detection in C-RANs. One interesting and promising topic is to extend the results in this book to the downlink beamforming problem. Unlike the uplink case, the beamforming design in downlink involves transmit power constraints of individual RRHs. This makes the downlink beamforming more complicated. Notice that a downlink beamforming problem may be efficiently transformed to an uplink detection problem. For example, [1] showed that the dual of a multi-antenna downlink channel with per-antenna power constraints is an uplink channel with noises having an uncertain diagonal covariance matrix. Then, the original downlink

problem can be solved by iteratively updating the noise covariance matrix and the dual uplink detection matrix. Thanks to the similarity between channel models of C-RANs and traditional multi-antenna systems, the downlink-uplink duality given in [1] also holds for C-RANs. This indicates that the downlink beamforming problem in C-RANs may be solved by existing efficient uplink algorithms, such as those proposed in this book.

2. **Resource Management:**
   Besides the full-scale coordinated signal processing, full-scale optimization of sharing, isolation, and customization of radio resources among coexisting mobile network operators and users becomes possible because of the centralized C-RAN architecture. The exciting opportunities come with unique technical challenges, among which one is particularly outstanding. That is, the cost and complexity of full-scale resource management is prohibitively high. Hence, it is critical to design scalable resource management schemes that can leverage the enormous centralized baseband-processing power to attain unprecedented spectrum efficiency at affordable cost and complexity. The successful signal processing algorithms proposed in this book show the possibility of designing scalable resource management algorithms by exploiting the near-sparsity of the channel matrix.
3. **Green Architecture:**
   The special features of C-RAN lead to unique architecture problems, such as:

   - *BBU Management:* The C-RAN architecture allows the data center to dynamically adjust the workload among BBUs. Efficient algorithms on computational power allocation and workload scheduling become a necessity. Moreover, notice that the wireless traffic load is highly dynamic in time and space. Thus, when the amount of workload is small, it is energy-efficient to consolidate all workload to a subset of BBUs, and turn off the idle BBUs. However, once in off state, a BBU cannot be restarted instantaneously. This would incur a service delay if the active BBUs cannot satisfy all the coming workload. Dynamic BBU management mechanisms are needed to balance the tradeoff between energy efficiency and service delay.
   - *Limited Capacity of Transport Network:* To guarantee a seamless cooperation among RRHs, data of RRHs are transmitting to the centralized data center from time to time. Even though high-bandwidth, low-latency optical transport links are employed in C-RANs, the capacity of the transport network is still finite. This limits the amount of information that can be exchanged between the BBUs and RRHs, which implies the cooperation among RRHs is limited. Moreover, the virtually centralized BBUs in the data center are geographically separated in general. Thus, the BBUs also need to share data over the transport network. How to fully utilize the limited backhaul capacity to maximize the cooperation and thereby optimize the system capacity is a critical problem in C-RANs.

- *RRH On/Off Problem:* Thanks to the high density of RRHs and fluctuation of wireless traffic load, not all the RRHs need to be active for all the time. For example, on weekends, it is essential to turn off some of RRHs in the office and industrial zones to decrease the power consumption in C-RANs. The corresponding power consumption of the transport network that are connected to these RRHs can also be reduced. The RRH on/off scheme is undoubtedly one of the key techniques to improve the energy efficiency in C-RANs.

Reference

1. W. Yu and T. Lan, "Transmitter optimization for the multi-antenna downlink with per-antenna power constraints," *IEEE Transactions on Signal Processing*, vol. 55, no. 6, pp. 2646–2660, 2007.

Index

A
Alternating direction method of multipliers (ADMM) algorithm, 72
A posteriori probability, 68
Approximate message passing (AMP), 72

B
Baseband units (BBUs), 2
BBU management, 95
Belief propagation, 67
Bipartite random geometric graph, 67–69
Blockwise randomized Gaussian message-passing (B-RGMP) algorithm, 6, 94
  computational complexity, 82
  computation time per iteration, 82
  convergence analysis, 83–84
  vs. RGMP, 85

C
Channel coefficients, 72
Channel matrix sparsification, 5
Channel sparsification, 10–11, 93
  GMP algorithm (*see* Gaussian message-passing algorithm)
  RGMP algorithm (*see* Randomized Gaussian message passing algorithm)
  signal detection, 94
Cloud radio access network (C-RAN)
  architecture, 2, 3, 68
  base station/antenna cooperation, 4
  channel matrix sparsification, 5
  channel sparsification, 93
  components, 2
  cost reduction, 2
  data-center technology, 3
  distributed antenna systems, 4
  low-complexity coordinated beamforming algorithm, 4
  multi-cell coordination schemes, 4
  normalized baseband processing complexity, 4
  PHY-layer collaborative signal processing, 3, 4
  RoF, 3
  scalable channel estimation, 5
  scalable signal detection, 5–6
  scalable signal processing algorithms, 93
  system capacity enhancement, 2
  system model, 9–10
  time-multiplexed channel training, 23
  training design scheme, 93
  uplink signal processing, 65, 94
  virtual BSs, 3
Computational complexity, 6
  B-RGMP algorithm, 82
  multi-layer DNC algorithm, 61–62
  RGMP algorithm, 73
  single-layer dynamic nested clustering, 51, 55–57

Y.-J. A. Zhang et al., *Scalable Signal Processing in Cloud Radio Access Networks*, SpringerBriefs in Electrical and Computer Engineering,
https://doi.org/10.1007/978-3-030-15884-2

Conjugate gradient (CG), 86, 87
Convergence analysis
  B-RGMP algorithm, 83–84
  comparison of, 85–86
  convergence speed, 86–88
  GMP algorithm, 77–79
  RGMP algorithm, 79–82
C-RAN, *see* Cloud radio access network

D
Damping factor, 86
Damping technique, 72
Data-center technology, 3
DBBD structure, *see* Doubly bordered block diagonal structure
Distance threshold analysis
  approximation, 12, 13
  vs. area radius, 19
  convergence, 19
  definition, 11
  distance distribution, 15
  eigenvalues, 15
  hypergeometric function, 16
  Jensen's inequality, 14
  K inequalities, 15–16
  lower bound of SINR ratio, 12
  non-zero entries, 17
  Pochhammer symbol, 16
  positive semidefinite matrix, 15
  vs. SINR ratio, 12, 13, 18–20
Distributed antenna systems, 4
Distributed remote radio heads, 2
DNC algorithm, *see* Dynamic nested clustering algorithm
Doubly bordered block diagonal (DBBD) structure, 5, 50, 51
Downlink beamforming, 94–95
Downlink-uplink duality, 95
DSATUR algorithm
  optimal training length, 36–37
  training sequence design, 32–33
Dynamic nested clustering (DNC) algorithm, 5–6, 94
  multi-layer DNC algorithm
    computational complexity, 61–62
    diagonal blocks, 58
    geographical RRH grouping, 58, 59
    light-grey boundary area, 60
    parallel computing, 62–63
    two-layer DNC algorithm, 60
    two-layer nested DBBD, 58, 59
  numerical results, 63–65
  single-layer dynamic nested clustering
    adjacent clusters, 52
    computational complexity, 51, 55–57
    DBBD matrix, 53–54
    disjoint clusters, 52
    iterative algorithms, 51
    non-iterative algorithm, 51
    parallel computing, 57–58
    RRH labelling algorithm, 52–54
  system model and problem formulation
    centralized BBU pool, 50
    channel matrix sparsification, 50
    DBBD matrix, 50, 51
    low-complexity algorithm, 50
    received signal vector at RRHs, 49
    transmitted signal vector, 50
    uplink transmission, 49

E
Equivalent interference-plus-noise, 44

G
Gaussian message-passing (GMP) algorithm, 94
  centralized data center, 69
  channel randomness, 71
  convergence of, 71, 73, 77–79
  Gaussian probability density functions, 70
  linear complexity per iteration, 71
  marginals, 69, 70
  related work, 71–72
  vs. RGMP, 85
  synchronous message passing, 73
  tree-type factor graph, 71
Gaussian probability density functions, 70
Generalized approximate message passing (GAMP), 72
General purpose processors (GPPs), 2
GMP algorithm, *see* Gaussian message-passing algorithm
Green architecture, 95–96

H
Hermitian DBBD matrix, 52
Hypergeometric function, 16

J
Jensen's inequality, 14

L

Large-scale fading coefficients, 25
Local orthogonality, 5, 93
  channel-estimation accuracy, 23
  channel sparsification, 27–29
  definition, 30
  graphical representation, 28, 29
  orthogonal training sequences, 27
  set of RRH indexes, 28
  throughput maximization, 23
Loop-free factor graphs, 72

M

Matrix stuffing, 72
Maximum *a posteriori* probability (MAP) detector, 68
Message passing
  B-RGMP algorithm (*see* Blockwise randomized Gaussian message-passing algorithm)
  GMP algorithm (*see* Gaussian message passing algorithm)
  RGMP algorithm (*see* Randomized Gaussian message passing algorithm)
  traditional message-passing algorithm, 67
Minimum mean square error (MMSE) detector, 3
Minimum training length, 23
MMSE detection algorithms, 5
Mobile devices, rapid proliferation, 1
Mobile networks
  challenges, 1–2
  densification, 1
  traffic demand, 1
Multi-cell coordination systems, 4
Multi-layer DNC algorithm
  computational complexity, 61–62
  diagonal blocks, 58
  geographical RRH grouping, 58, 59
  light-grey boundary area, 60
  parallel computing, 62–63
  two-layer DNC algorithm, 60
  two-layer nested DBBD, 58, 59
Multiuser multiple-input multiple-output (MU-MIMO) system, 71–72

O

Optimal training length
  asymptotic behavior of, 34–36
  chromatic number, 36, 37
  DSATUR algorithm, 36–37
  graph with infinite RRHs, 34
  minimum training length, 33, 37
  number of users, 37
  suboptimal coloring algorithm, 37
Orthogonal training sequences, 23

P

Parallel computing, 94
  multi-layer DNC algorithm, 62–63
  single-layer dynamic nested clustering, 57–58
Parallel signal processing, 57
Pochhammer symbol, 16

R

Radio access network (RAN) architecture, 1
Radio over fiber (RoF), 3
Random geometric graph theory, 5, 23
Randomized Gaussian message passing (RGMP) algorithm, 6, 74, 94
  B-RGMP algorithm, 82–85
  computation time of, 73
  continuous random variables, 73
  convergence of, 79–82
  vs. disjoint clustering algorithm, 88–89
  non-negligible computational complexity, 73
  numerical examples, 74–76
Rayleigh fading coefficient, 10
Refined channel sparsification, 38–39
Remote radio heads (RRHs), 2
  density, 93
  geographical grouping, 58, 59
  graph with infinite, 34
  and K users, 24
  labelling algorithm, 52–54
  on/off problem, 96
  set of RRH indexes, 28
Resource management, 95
RGMP algorithm, *see* Randomized Gaussian message passing algorithm
RRHs, *see* Remote radio heads

S

Scalable channel estimation, 5
  local orthogonality
    channel sparsification, 27–29
    definition, 30
    graphical representation, 28, 29
    orthogonal training sequences, 27
    set of RRH indexes, 28

Scalable channel estimation (*cont.*)
  optimal training length
    asymptotic behavior of, 34–36
    chromatic number, 36, 37
    DSATUR algorithm, 36–37
    graph with infinite RRHs, 34
    minimum training length, 33, 37
    number of users, 37
    suboptimal coloring algorithm, 37
  problem statement, 30–31
  related work, 31
  system model
    data transmission phase, 26
    large-scale fading coefficients, 25
    power constraint, 24
    RRHs and K users, 24
    signal model, 24
    small-scale fading coefficients, 25
    training phase, 25–26
  throughput optimization, 26–27
  training-based CRAN scheme
    information throughput, 38
    numerical results, 39–43
    refined channel sparsification, 38–39
    throughput lower bound, 43–45
  training sequence design, 32–33
Scalable collaborative signal processing algorithms, 4
Scalable signal detection, 5–6
  bipartite random geometric graph, 67–69
  DNC algorithm (*see* Dynamic nested clustering algorithm)
  GMP algorithm (*see* Gaussian message-passing algorithm)
  RGMP algorithm (*see* Randomized Gaussian message passing algorithm)
Scalable training-based C-RAN design, 23
Signal processing
  in MU-MIMO systems, 71
  parallel, 57
  PHY-layer collaborative signal processing, 3, 4
  scalable signal processing algorithms, 4, 93
  uplink, 65, 94
Signal-to-interference-and-noise ratio (SINR)
  loss, 5
  for user k, 10
Single-layer dynamic nested clustering
  adjacent clusters, 52
  computational complexity, 51, 55–57
  DBBD matrix, 53–54
  disjoint clusters, 52
  iterative algorithms, 51
  non-iterative algorithm, 51
  parallel computing, 57–58
  RRH labelling algorithm, 52–54
SINR ratio
  distance threshold, 12–16, 18–20
  lower bound of, 12
Small-scale fading coefficients, 25
Sparsified channel matrix, 5
Synchronous GMP, 73

T

Threshold-based channel matrix sparsification method, 5
Training-based CRAN scheme
  information throughput, 38
  numerical results, 39–43
  refined channel sparsification, 38–39
  throughput lower bound, 43–45
Transport network, limited capacity of, 95
Tree-type factor graph, 71
Truncated heavy-tailed distribution, 72
Two-layer DNC algorithm, 60
Two-layer nested DBBD, 58, 59

V

Vertex-coloring algorithms, 23
Vertex-coloring problem, 5

Printed in the United States
By Bookmasters